Reasonable Books = Reasonable Prices

Irina Kleyman

Exceptional for Home Schooling

Math Help Grade 3

AF875445

First Edition

Reasonable Books = Reasonable Prices

Irina Kleyman

Exceptional for Home Schooling

Math Help
Grade 3

I dedicate this book to my children and grandchildren.

Acknowledgments
Special thanks to Michael G LaFratta, Vladimir Kleyman,
who has assisted in bringing all 10 books to market.

Author's note: In the many years that I've taught mathematics, I have been informed by many of my students that my methods for teaching have helped them tremendously in learning math. I have written these books in order to continue to help others learn math effectively.

Math Help Grade 1
Math Help Grade 2
Math Help Grade 3
Math Help Grade 4
Math Help Grade 5
Math Help Grade 6
Math Help Grade 7
Math Help Grade 8
Math Help for College (Summary Grades 1 – 6)
Math Help Grade for University (Basic, Algebra, Geometry)

Copyright © 0169 2020
ISBN: 978-1-7349699-2-4 Paperback

All rights reserved. No part of this book may be reproduced or transmitted in any form or by any means, electronic or mechanical, including photocopying, recording, or by any information storage and retrieval system, without permission in writing from the copyright owner.

Printed in the United States of America.

Syllabus Math 1, 2, 3, 4

Grade 1 # 1 – 4 Grade 2 # 5 – 8 Grade 3 # 9-12 Grade 4 # 13 – 16

Syllabus includes topics, what students need to know.
Unknowns are numbers, which we need to find.

#	Grades	Topics. Advanced Problems (A-4), Pages (p)
	Chapters	Basic Problems (B), Chapter (C), Lesson (L)
1	1-1	One-Digit Numbers (B)
2	1-2	Two-Digit Numbers (B)
3	1-3	Three numbers (B)
4	1-4	Finding unknowns (A-1)
	Grade 1	Final Exam 1 B, C 4, L 12 # 1-20 odd, p 94
5	2-1	Addition or Subtraction Numbers (20–29) (B)
6	2-2	Addition or Subtraction Numbers (50–59) (B)
7	2-3	Addition or Subtraction Numbers (30–39) (B)
8	2-4	Finding unknowns (A-2)
	Grade 2	Final Exam 2 B, C 4, L 17, # 1 – 20 odd, p 106
9	3-1	Addition or Subtraction Numbers (60–69) (B)
10	3-2	Addition or Subtraction Numbers (90–99) (B)
11	3-3	Addition or Subtraction Numbers (70–79) (B)
12	3-4	Finding unknowns (A-3)
	Grade 3	Final Exam 3 B, C 4, L 19, # 1 – 20 odd, p 121
13	4-1	Three-Digit numbers (B)
14	4-2	Multiplication numbers (B)
15	4-3	Division numbers (B)
16	4-4	Finding unknowns (A-4)
	Grade 4	Final Exam 4 B, C 4, L 12, #1-10 all, p 133

Parents, Tutors, Students Middle School, High School, Colleges, University can help primary school students. Please read slowly math books to primary school students.

Grade 3

www.mathbooksbyirina.com

Chapter 1 Addition or Subtraction Numbers (60-69)

Grade 3 Chapter 1 Contents
Numbers 60 – 69
Lesson 1 - 11
Basic Problems

Work with lesson 1, read rules and examples, 2) Solve the problems from contents, 3) Check answers in the end of chapter 1, 4) Work with the next lessons 2, 3, 4, 5, …, 5) Solve Test 1 A, 6) Grading Scale: A =10 points, B=9points, C=8 points, D=7 points, F=0-6 points, 7) If you solve 10 problems correctly, you get an A, 8) If have an A, you can move on to next chapter, 9) If your grade is lower than an A, work with the mistakes, then solve Test 1B, 10) When you finish all of the chapters, solve Final Exam in last chapter.

Grade 3 Chapter 1 Lesson 1
Review 1 A
Add or Subtract numbers in row

Add

1) 9 + 1
2) 8 + 2
3) 7 + 3
4) 6 + 4
5) 5 + 5
6) 9 + 2
7) 8 + 3
8) 7 + 4
9) 6 + 5
10) 9 + 3
11) 8 + 4
12) 7 + 5
13) 6 + 6
14) 9 + 4
15) 8 + 5
16) 7 + 6
17) 9 + 5
18) 8 + 6
19) 7 + 7
20) 9 + 6
21) 8 + 7
22) 9 + 7
23) 8 + 8
24) 9 + 8
25) 9 + 9

Subtract

1) 10 – 1
2) 10 – 2
3) 10 – 3
4) 10 – 4
5) 10 – 5
6) 11 - 2
7) 11 – 3
8) 11 – 4
9) 11 – 5
10) 12 – 3
11) 12 – 4
12) 12 – 5
13) 12 – 6
14) 13 – 4
15) 13 – 5
16) 13 – 6
17) 14 – 5
18) 14 – 6
19) 14 – 7
20) 15 – 6
21) 15 – 7
22) 16 – 7
23) 16 – 8
24) 17 – 8
25) 18 - 9

Answers Lesson 1 Review 1 A

Add 1) 10 2) 10 3) 10 4) 10 5) 10 6) 11 7) 11 8) 11 9) 11 10) 12 11) 12 12) 12 13) 12 14) 13 15) 13 16) 13 17) 14 18) 14 19) 14 20) 15 21) 15 22) 16 23) 16 24) 17 25) 18

Subtract 1) 9 2) 8 3) 7 4) 6 5) 5 6) 9 7) 8 8) 7 9) 6 10) 9 11) 8 12) 7 13) 6 14) 9 15) 8 16) 7 17) 9 18) 8 19) 7 20) 9 21) 8 22) 9 23) 8 24) 9 25) 9

Grade 3 Chapter 1 Lesson 1
Review 1 B
Add or Subtract numbers in row

Number 20
6 + 14 = 14 + 6 =10+4+6=10+10= 20
20 – 14 = 20 – 10 – 4 = 10 – 4 = 6
Number 21
7 + 14 = 14 + 7 = 10 + 4 + 7 =10 + 11 = 21
21 – 14 = 21 – 11 – 3 = 10 – 3 = 7
Number 22
4 + 18 = 18 + 4 = 10 + 8 + 4 = 10 + 12 = 22
22 – 14 = 22 – 12 – 2 = 10 – 2 = 8
Number 23
9 + 14 = 14 + 9 = 10 + 4 + 9=10+13 = 23
23 – 14 = 23 – 13 – 1 = 10 – 1 = 9
Number 24
10 + 14 = 10 + 10 + 4 = 20 + 4 = 24
24 – 14 = 24 – 10 – 4 = 14 – 4 = 10
Number 25
11 + 14 = 10 + 1 + 10 + 4 = 20 + 5 = 25
25 – 14 = 25 – 10 – 4 = 15 – 4 = 11
Numbers 26
12 + 14 = 10 + 2 + 10 + 4 = 20 + 6 = 26
26 - 14 = 26 – 10 – 4 = 16 – 4 = 12
Number 27
13 + 14 = 10 + 3 + 10 + 4 = 20 + 7 = 27
27 – 14 = 27 – 10 – 4 = 17 – 4 = 13
Number 28
14 + 14 = 10 + 4 + 10 + 4 = 20 + 8 = 28
28 – 14 = 28 – 10 – 4 = 18 – 4 = 14
Number 29
15 + 14 = 10 + 5 + 10 + 4 = 20 + 9 = 29
29 – 14 = 29 – 10 – 4 = 19 – 4 = 15

Grade 3 Chapter 1 Lesson 2
Review 1 A
Add or Subtract numbers in column

5 tens – 1 ten = 4 tens   10 ones = 1 ten

1) Number 50

a)	b)	c)	d)
1	4 10	1	4 10
29	5 0	19	50
+21	- 2 1	+31	- 19
50	2 9	50	31

2) Number 51

a)	b)	c)	d)
1	4 11	1	4 11
29	5 1	19	51
+22	- 2 2	+32	- 19
51	2 9	51	32

3) Number 52

a)	b)	c)	d)
1	4 12	1	4 12
29	5 2	19	52
+23	- 2 3	+33	- 19
52	2 9	52	33

4) Number 53

a)	b)	c)	d)
1	4 13	1	4 13
29	5 3	19	53
+24	- 2 4	+34	- 19
53	2 9	53	34

5) Numbers 54

1)	2)	3)	4)
1	4 14	1	4 14
29	5 4	19	54
+25	- 2 5	+ 35	- 19
54	2 9	54	35

Grade 3 Chapter 1 Lesson 2
Review 1 B
Add or Subtract numbers in column

5 tens – 1 ten = 4 tens   10 ones = 1 ten

6) Numbers 55

a)	b)	c)	d)
$\begin{array}{r} ^{1} \\ 29 \\ +26 \\ \hline 55 \end{array}$	$\begin{array}{r} ^{4\ 15} \\ 5\ 5 \\ -\ 2\ 6 \\ \hline 2\ 9 \end{array}$	$\begin{array}{r} ^{1} \\ 19 \\ +36 \\ \hline 55 \end{array}$	$\begin{array}{r} ^{4\ 15} \\ 55 \\ -\ 19 \\ \hline 36 \end{array}$

7) Numbers 56

a)	b)	c)	d)
$\begin{array}{r} ^{1} \\ 29 \\ +27 \\ \hline 56 \end{array}$	$\begin{array}{r} ^{4\ 16} \\ 5\ 6 \\ -\ 2\ 7 \\ \hline 2\ 9 \end{array}$	$\begin{array}{r} ^{1} \\ 19 \\ +37 \\ \hline 56 \end{array}$	$\begin{array}{r} ^{4\ 16} \\ 56 \\ -\ 19 \\ \hline 37 \end{array}$

8) Numbers 57

a)	b)	c)	d)
$\begin{array}{r} ^{1} \\ 29 \\ +28 \\ \hline 57 \end{array}$	$\begin{array}{r} ^{4\ 17} \\ 5\ 7 \\ -\ 2\ 8 \\ \hline 2\ 9 \end{array}$	$\begin{array}{r} ^{1} \\ 19 \\ +38 \\ \hline 57 \end{array}$	$\begin{array}{r} ^{4\ 17} \\ 57 \\ -\ 19 \\ \hline 38 \end{array}$

9) Numbers 58

a)	b)	c)	d)
$\begin{array}{r} ^{1} \\ 29 \\ +29 \\ \hline 58 \end{array}$	$\begin{array}{r} ^{4\ 18} \\ 5\ 8 \\ -\ 2\ 9 \\ \hline 2\ 9 \end{array}$	$\begin{array}{r} ^{1} \\ 19 \\ +39 \\ \hline 58 \end{array}$	$\begin{array}{r} ^{4\ 18} \\ 58 \\ -\ 19 \\ \hline 39 \end{array}$

10) Numbers 59

a)	b)	c)	d)
$\begin{array}{r} 29 \\ +\ 30 \\ \hline 59 \end{array}$	$\begin{array}{r} 59 \\ -\ 30 \\ \hline 29 \end{array}$	$\begin{array}{r} 19 \\ +40 \\ \hline 59 \end{array}$	$\begin{array}{r} 59 \\ -19 \\ \hline 40 \end{array}$

Grade 3 Chapter 1 Lesson 3
Review 1 A
Add or Subtract Numbers in Row

20 – 16 = 20 – 10 – 6 = 10 – 6 = 4

# 1	31 + 19
# 2	29 + 22
# 3	13 + 39
# 4	29 + 24
# 5	25 + 29
# 6	20 – 11
# 7	20 – 12
# 8	20 – 13
# 9	20 – 14
# 10	20 – 15

Add or Subtract Numbers in Column

# 11	# 12	# 13	# 14	# 15
26	29	28	19	40
+29	+27	+ 29	+39	+ 19

# 16	# 17	# 18	# 19	# 20
55	56	57	58	59
-29	-27	-29	-39	- 19

Answers Lesson 3 Review 1 A

# 1	50	# 11	55
# 2	51	# 12	56
# 3	52	# 13	57
# 4	53	# 14	58
# 5	54	# 15	59
# 6	9	# 16	26
# 7	8	# 17	29
# 8	7	# 18	28
# 9	6	# 19	19
# 10	5	# 20	40

Grade 3 Chapter 1 Lesson 3
Review 1 B
Add or Subtract Numbers in Row

30 – 16 = 30 – 10 – 6 = 20 – 6 = 14

# 1	11 + 39
# 2	29 + 22
# 3	23 + 29
# 4	29 + 34
# 5	15 + 39
# 6	30 – 11
# 7	30 – 12
# 8	30 – 13
# 9	30 – 14
# 10	30 – 15

Add or Subtract Numbers in Column

# 11 29 +26	# 12 27 +29	# 13 29 + 27	# 14 39 +19	# 15 19 + 40
# 16 55 -26	# 17 56 -29	# 18 57 -18	# 19 58 -39	# 20 59 - 19

Answers Lesson 3 Review 1 B

# 1	50	# 11	55
# 2	51	# 12	56
# 3	52	# 13	56
# 4	63	# 14	58
# 5	54	# 15	59
# 6	19	# 16	29
# 7	18	# 17	27
# 8	17	# 18	39
# 9	16	# 19	19
# 10	15	# 20	40

Grade 3 Chapter 1 Lesson 4

Add and Subtract three numbers

Three numbers add in any order

11 + 14 + 19 = 11 + 19 + 14 = 30 + 14 = 44

11 + 14 + 19 = 25 + 19 = 44

11 + 14 + 19 = 11 + 33 = 44

Review 1 A

Add three numbers

	a	b	c	d
# 1	20 + 0 + 9	30 + 0 + 9	40 + 0 + 9	50 + 0 + 9
# 2	21 + 1 + 8	31 + 1 + 8	41 + 1 + 8	51 + 1 + 8
# 3	22 + 2 + 7	32 + 2 + 7	42 + 2 + 7	52 + 2 + 7
# 4	23 + 3 + 6	33 + 3 + 6	43 + 3 + 6	53 + 3 + 6
# 5	24 + 4 + 5	34 + 4 + 5	44 + 4 + 5	54 + 4 + 5
# 6	25 + 5 + 4	35 + 5 + 4	45 + 5 + 4	55 + 5 + 4
# 7	26 + 6 + 3	36 + 6 + 3	46 + 6 + 3	56 + 6 + 3
# 8	27 + 7 + 2	37 + 7 + 2	47 + 7 + 2	57 + 7 + 2
# 9	28 + 8 + 1	38 + 8 + 1	48 + 8 + 1	58 + 8 + 1
# 10	29 + 9 + 0	39 + 9 + 0	49 + 9 + 0	59 + 9 + 0

Answers Lesson 4 Review 1 A

	a	b	c	d
# 1	29	39	49	59
# 2	30	40	50	60
# 3	31	41	51	61
# 4	32	42	52	62
# 5	33	43	53	63
# 6	34	44	54	64
# 7	35	45	55	65
# 8	36	46	56	66
# 9	37	47	57	67
# 10	38	48	58	68

Grade 3 Chapter 1 Lesson 4

Add and Subtract three numbers

To subtract three numbers, work from left to right

19 – 1 – 9 = 18 – 9 = 9

18 – 2 – 8 = 16 – 8 = 8

18 – 8 – 2 = 10 – 2 = 8

Review 1 B

Subtract three numbers

	a	b	c	d
# 11	60 – 9 – 0	70 – 9 – 0	80 – 9 – 0	90 – 9 – 0
# 12	61 – 8 – 1	71 – 8 – 1	81 – 8 – 1	91 – 8 – 1
# 13	62 – 7 – 2	72 – 7 – 2	82 - 7 – 2	92 – 7 – 2
# 14	63 – 6 – 3	73 – 6 – 3	83 – 6 – 3	93 – 6 – 3
# 15	64 – 5 – 4	74 – 5 – 4	84 – 5 – 4	94 – 5 – 4
# 16	65 – 4 – 5	75 – 4 - 5	85 - 4 – 5	95 – 4 – 5
# 17	66 – 3 – 6	76 – 3 – 6	86 – 3 – 6	96 – 3 – 6
# 18	67 – 2 – 7	77 – 2 – 7	87 – 2 – 7	97 – 2 – 7
# 19	68 – 1 – 8	78 – 1 – 8	88 – 1 – 8	98 – 1 – 8
# 20	69 – 0 – 9	79 – 0 – 9	89 – 0 – 9	99 – 0 - 9

Answers Lesson 4 Review 1 B

	a	b	c	d
# 11	51	61	71	81
# 12	52	62	72	82
# 13	53	63	73	83
# 14	54	64	74	84
# 15	55	65	75	85
# 16	56	66	76	86
# 17	57	67	77	87
# 18	58	68	78	88
# 19	59	69	79	89
# 20	60	70	80	90

Grade 3 Chapter 1 Lesson 5

Two operations with numbers

To do two operations work from left to right

19 + 9 - 1 = 28 - 1 = 27

18 + 8 - 2 = 26 - 2 = 24

Review 1 A

Add or Subtract numbers

	a	b	c	d
# 1	20 + 9 – 0	30 + 9 – 0	40 + 9 – 0	50 + 9 - 0
# 2	21 + 8 – 1	31 + 8 – 1	41 + 8 – 1	51 + 8 - 1
# 3	22 + 7 – 2	32 + 7 – 2	42 + 7 – 2	52 + 7 - 2
# 4	23 + 6 – 3	33 + 6 – 3	43 + 6 – 3	53 + 6 - 3
# 5	24 + 5 – 4	34 + 5 – 4	44 + 5 – 4	54 + 5 - 4
# 6	25 + 4 – 5	35 + 4 – 5	45 + 4 – 5	55 + 4 - 5
# 7	26 + 3 – 6	36 + 3 – 6	46 + 3 – 6	56 + 3 - 6
# 8	27 + 2 – 7	37 + 2 – 7	47 + 2 – 7	57 + 2 - 7
# 9	28 + 1 – 8	38 + 1 – 8	48 + 1 – 8	58 + 1 - 8
# 10	29 + 0 – 9	39 + 0 – 9	49 + 0 – 9	59 + 0 - 9

Answers Lesson 5 Review 1 A

	a	b	c	d
# 1	29	39	49	59
# 2	28	38	48	58
# 3	27	37	47	57
# 4	26	36	46	56
# 5	25	35	45	55
# 6	24	34	44	54
# 7	23	33	43	53
# 8	22	32	42	52
# 9	21	31	41	51
# 10	20	30	40	50

Grade 3 Chapter 1 Lesson 5

Two operations with numbers

To do two operations work from left to right

19 - 8 + 9 = 11 + 9 = 20

18 - 9 + 8 = 9 + 8 = 17

Review 1 B

Subtract or Add Numbers

	a	b	c	d
# 11	20 - 9 + 0	30 - 9 + 0	40 - 9 + 0	50 - 9 + 0
# 12	21 - 8 + 1	31 - 8 + 1	41 - 8 + 1	51 - 8 + 1
# 13	22 - 7 + 2	32 - 7 + 2	42 - 7 + 2	52 - 7 + 2
# 14	23 - 6 + 3	33 - 6 + 3	43 - 6 + 3	53 - 6 + 3
# 15	24 - 5 + 4	34 - 5 + 4	44 - 5 + 4	54 - 5 + 4
# 16	25 - 4 + 5	35 - 4 + 5	45 - 4 + 5	55 - 4 + 5
# 17	26 - 3 + 6	36 - 3 + 6	46 - 3 + 6	56 - 3 + 6
# 18	27 - 2 + 7	37 - 2 + 7	47 - 2 + 7	57 - 2 + 7
# 19	28 - 1 + 8	38 - 1 + 8	48 - 1 + 8	58 - 1 + 8
# 20	29 - 0 + 9	39 - 0 + 9	49 - 0 + 9	59 - 0 + 9

Answers Lesson 5 Review 1 B

	a	b	c	d
# 11	11	21	31	41
# 12	14	24	34	44
# 13	17	27	37	47
# 14	20	30	40	50
# 15	23	33	43	53
# 16	26	36	46	56
# 17	29	39	49	59
# 18	32	42	52	62
# 19	35	45	55	65
# 20	38	48	58	68

Grade 3 Chapter 1 Lesson 6
Number 60

31 + 29 =30 + 1 + 20 + 9 = 30 + 20+ 1 + 9 = 50 + 10 = 60

60 – 29 = 60 – 20 – 9 = 40 – 9 = 31

22 + 38 = 20 + 2 + 30 + 8 = 20 + 30 + 2 + 8 = 50 + 10 = 60

60 – 38 = 60 – 30 – 8 = 30 – 8 = 22

Add or Subtract numbers in row

	a	b	c	d
# 1	60 – 0	40 + 20	60 – 20	60 - 40
# 2	60 – 1	39 + 21	60 – 21	60 - 41
# 3	60 – 2	38 + 22	60 – 22	60 - 42
# 4	60 – 3	37 + 23	60 – 23	60 - 43
# 5	60 – 4	36 + 24	60 – 24	60 - 44
# 6	60 – 5	35 + 25	60 – 25	60 - 45
# 7	60 – 6	34 + 26	60 – 26	60 - 46
# 8	60 – 7	33 + 27	60 – 27	60 - 47
# 9	60 – 8	32 + 28	60 – 28	60 - 48
# 10	60 – 9	31 + 29	60 – 29	60 - 49
# 11	60 – 10	30 + 30	60 – 30	60 – 50
# 12	60 – 11	29 + 31	60 – 31	60 - 51
# 13	60 – 12	28 + 32	60 – 32	60 – 52
# 14	60 – 13	27 + 33	60 – 33	60 - 53
# 15	60 – 14	26 + 34	60 – 34	60 – 54
# 16	60 – 15	25 + 35	60 – 35	60 - 55
# 17	60 – 16	24 + 36	60 – 36	60 – 56
# 18	60 – 17	23 + 37	60 – 37	60 – 57
# 19	60 – 18	22 + 38	60 – 38	60 - 58
# 20	60 – 19	21 + 39	60 – 39	60 – 59

Grade 3 Chapter 1 Lesson 6
Number 61

32 + 29 =30 + 2 + 20 + 9 = 30 + 20 + 2 + 9 = 50 + 11 = 61
61 – 29 = 61 – 20 – 9 = 41 – 9 = 32
22 + 39 = 20 + 2 + 30 + 9 = 20 + 30 + 2 + 9 = 50 + 11 = 61
61 – 39 = 61 – 30 – 9 = 31 – 9 = 22

Add or Subtract numbers in row

	a	b	c	d
# 1	61 – 0	41 + 20	61 – 20	61 - 40
# 2	61 – 1	40 + 21	61 – 21	61 - 41
# 3	61 – 2	39 + 22	61 – 22	61 - 42
# 4	61 – 3	38 + 23	61 – 23	61 - 43
# 5	61 – 4	37 + 24	61 – 24	61 - 44
# 6	61 – 5	36 + 25	61 – 25	61 - 45
# 7	61 – 6	35 + 26	61 – 26	61 - 46
# 8	61 – 7	34 + 27	61 – 27	61 - 47
# 9	61 – 8	33 + 28	61 – 28	61 - 48
# 10	61 – 9	32 + 29	61 – 29	61 - 49
# 11	61 – 10	31 + 30	61 – 30	61 – 50
# 12	61 – 11	30 + 31	61 – 31	61 - 51
# 13	61 – 12	29 + 32	61 – 32	61 – 52
# 14	61 – 13	28 + 33	61 – 33	61 - 53
# 15	61 – 14	27 + 34	61 – 34	61 – 54
# 16	61 – 15	26 + 35	61 – 35	61 - 55
# 17	61 – 16	25 + 36	61 – 36	61 – 56
# 18	61 – 17	24 + 37	61 – 37	61 – 57
# 19	61 – 18	23 + 38	61 – 38	61 - 58
# 20	61 – 19	22 + 39	61 – 39	61 – 59

Grade 3 Chapter 1 Lesson 7

Number 62

33 + 29 =30 + 3 + 20 + 9 = 30 + 20 + 3 + 9 = 50 + 12 = 62

62 – 29 = 62 – 20 – 9 = 42 – 9 = 33

23 + 39 = 20 + 3 + 30 + 9 = 20 + 30 + 3 + 9 = 50 + 12 = 62

62 – 39 = 62 – 30 – 9 = 32 – 9 = 23

Add or Subtract numbers in row

	a	b	c	d
# 1	62 – 0	42 + 20	62 – 20	62 - 40
# 2	62 – 1	41 + 21	62 – 21	62 - 41
# 3	62 – 2	40 + 22	62 – 22	62 - 42
# 4	62 – 3	39 + 23	62 – 23	62 - 43
# 5	62 – 4	38 + 24	62 – 24	62 - 44
# 6	62 – 5	37 + 25	62 – 25	62 - 45
# 7	62 – 6	36 + 26	62 – 26	62 - 46
# 8	62 – 7	35 + 27	62 – 27	62 - 47
# 9	62 – 8	34 + 28	62 – 28	62 - 48
# 10	62 – 9	33 + 29	62 – 29	62 - 49
# 11	62 – 10	32 + 30	62 – 30	62 – 50
# 12	62 – 11	31 + 31	62 – 31	62 - 51
# 13	62 – 12	30 + 32	62 – 32	62 – 52
# 14	62 – 13	29 + 33	62 – 33	62 - 53
# 15	62 – 14	28 + 34	62 – 34	62 – 54
# 16	62 – 15	27 + 35	62 – 35	62 - 55
# 17	62 – 16	26 + 36	62 – 36	62 – 56
# 18	62 – 17	25 + 37	62 – 37	62 – 57
# 19	62 – 18	24 + 38	62 – 38	62 - 58
# 20	62 – 19	23 + 39	62 – 39	62 – 59

Grade 3 Chapter 1 Lesson 7
Number 63

34 + 29 =30 + 4 + 20 + 9 = 30 + 20 + 4 + 9 = 50 + 13 = 63

63 – 29 = 63 – 20 – 9 = 43 – 9 = 34

24 + 39 = 20 + 4 + 30 + 9 = 20 + 30 + 4 + 9 = 50 + 13 = 63

63 – 39 = 63 – 30 – 9 = 33 – 9 = 24

Add or Subtract numbers in row

	a	b	c	d
# 1	63 – 0	43 + 20	63 – 20	63 - 40
# 2	63 – 1	42 + 21	63 – 21	63 - 41
# 3	63 – 2	41 + 22	63 – 22	63 - 42
# 4	63 – 3	40 + 23	63 – 23	63 - 43
# 5	63 – 4	39 + 24	63 – 24	63 - 44
# 6	63 – 5	38 + 25	63 – 25	63 - 45
# 7	63 – 6	37 + 26	63 – 26	63 - 46
# 8	63 – 7	36 + 27	63 – 27	63 - 47
# 9	63 – 8	35 + 28	63 – 28	63 - 48
# 10	63 – 9	34 + 29	63 – 29	63 - 49
# 11	63 – 10	33 + 30	63 – 30	63 – 50
# 12	63 – 11	32 + 31	63 – 31	63 - 51
# 13	63 – 12	31 + 32	63 – 32	63 – 52
# 14	63 – 13	30 + 33	63 – 33	63 - 53
# 15	63 – 14	29 + 34	63 – 34	63 – 54
# 16	63 – 15	28 + 35	63 – 35	63 - 55
# 17	63 – 16	27 + 36	63 – 36	63 – 56
# 18	63 – 17	26 + 37	63 – 37	63 – 57
# 19	63 – 18	25 + 38	63 – 38	63 - 58
# 20	63 – 19	24 + 39	63 – 39	63 – 59

Grade 3 Chapter 1 Lesson 8
Number 64

35 + 29 =30 + 5 + 20 + 9 = 30 + 20 + 5 + 9 = 50 + 14 = 64
64 – 29 = 64 – 20 – 9 = 44 – 9 = 35
25 + 39 = 20 + 5 + 30 + 9 = 20 + 30 + 5 + 9 = 50 + 14 = 64
64 – 39 = 64 – 30 – 9 = 34 – 9 = 25

Add or Subtract numbers in row

	a	b	c	d
# 1	64 – 0	44 + 20	64 – 20	64 - 40
# 2	64 – 1	43 + 21	64 – 21	64 - 41
# 3	64 – 2	42 + 22	64 – 22	64 - 42
# 4	64 – 3	41 + 23	64 – 23	64 - 43
# 5	64 – 4	40 + 24	64 – 24	64 - 44
# 6	64 – 5	39 + 25	64 – 25	64 - 45
# 7	64 – 6	38 + 26	64 – 26	64 - 46
# 8	64 – 7	37 + 27	64 – 27	64 - 47
# 9	64 – 8	36 + 28	64 – 28	64 - 48
# 10	64 – 9	35 + 29	64 – 29	64 - 49
# 11	64 – 10	34 + 30	64 – 30	64 – 50
# 12	64 – 11	33 + 31	64 – 31	64 - 51
# 13	64 – 12	32 + 32	64 – 32	64 – 52
# 14	64 – 13	31 + 33	64 – 33	64 - 53
# 15	64 – 14	30 + 34	64 – 34	64 – 54
# 16	64 – 15	29 + 35	64 – 35	64 - 55
# 17	64 – 16	28 + 36	64 – 36	64 – 56
# 18	64 – 17	27 + 37	64 – 37	64 – 57
# 19	64 – 18	26 + 38	64 – 38	64 - 58
# 20	64 – 19	25 + 39	64 – 39	64 – 59

Grade 3 Chapter 1 Lesson 8
Number 65

36 + 29 =30 + 6 + 20 + 9 = 30 + 20 + 6 + 9 = 50 + 15 = 65

65 – 29 = 65 – 20 – 9 = 45 – 9 = 36

26 + 39 = 20 + 6 + 30 + 9 = 20 + 30 + 6 + 9 = 50 + 15 = 65

65 – 39 = 65 – 30 – 9 = 35 – 9 = 26

Add or Subtract numbers in row

	a	b	c	d
# 1	65 – 0	45 + 20	65 – 20	65 - 40
# 2	65 - 1	44 + 21	65 – 21	65 - 41
# 3	65 – 2	43 + 22	65 – 22	65 - 42
# 4	65 - 3	42 + 23	65 – 23	65 - 43
# 5	65 – 4	41 + 24	65 – 24	65 - 44
# 6	65 – 5	40 + 25	65 – 25	65 - 45
# 7	65 – 6	39 + 26	65 – 26	65 - 46
# 8	65 – 7	38 + 27	65 – 27	65 - 47
# 9	65 – 8	37 + 28	65 – 28	65 - 48
# 10	65 – 9	36 + 29	65 – 29	65 - 49
# 11	65 – 10	35 + 30	65 – 30	65 – 50
# 12	65 – 11	34 + 31	65 – 31	65 - 51
# 13	65 – 12	33 + 32	65 – 32	65 – 52
# 14	65 – 13	32 + 33	65 – 33	65 - 53
# 15	65 – 14	31 + 34	65 – 34	65 – 54
# 16	65 – 15	30 + 35	65 – 35	65 - 55
# 17	65 – 16	29 + 36	65 – 36	65 – 56
# 18	65 – 17	28 + 37	65 – 37	65 – 57
# 19	65 – 18	27 + 38	65 – 38	65 - 58
# 20	65 – 19	26 + 39	65 – 39	65 – 59

Grade 3 Chapter 1 Lesson 9

Number 66

37 + 29 =30 + 7 + 20 + 9 = 30 + 20 + 7 + 9 = 50 + 16 = 66

66 – 29 = 66 – 20 – 9 = 46 – 9 = 37

27 + 39 = 20 + 7 + 30 + 9 = 20 + 30 + 7 + 9 = 50 + 16 = 66

66 – 39 = 66 – 30 – 9 = 36 – 9 = 27

Add or Subtract Numbers in row

	a	b	c	d
# 1	66 – 0	46 + 20	66 – 20	66 - 40
# 2	66 – 1	45 + 21	66 – 21	66 - 41
# 3	66 – 2	44 + 22	66 – 22	66 - 42
# 4	66 – 3	43 + 23	66 – 23	66 - 43
# 5	66 – 4	42 + 24	66 – 24	66 - 44
# 6	66 – 5	41 + 25	66 – 25	66 - 45
# 7	66 – 6	40 + 26	66 – 26	66 - 46
# 8	66 – 7	39 + 27	66 – 27	66 - 47
# 9	66 – 8	38 + 28	66 – 28	66 - 48
# 10	66 – 9	37 + 29	66 – 29	66 - 49
# 11	66 – 10	36 + 30	66 – 30	66 – 50
# 12	66 – 11	35 + 31	66 – 31	66 - 51
# 13	66 – 12	34 + 32	66 – 32	66 – 52
# 14	66 – 13	33 + 33	66 – 33	66 - 53
# 15	66 – 14	32 + 34	66 – 34	66 – 54
# 16	66 – 15	31 + 35	66 – 35	66 - 55
# 17	66 – 16	30 + 36	66 – 36	66 – 56
# 18	66 – 17	29 + 37	66 – 37	66 – 57
# 19	66 – 18	28 + 38	66 – 38	66 - 58
# 20	66 – 19	27 + 39	66 – 39	66 – 59

Grade 3 Chapter 1 Lesson 9

Number 67

38 + 29 =30 + 8 + 20 + 9 = 30 + 20 + 8 + 9 = 50 + 17 = 67

67 – 29 = 67 – 20 – 9 = 47 – 9 = 38

28 + 39 = 20 + 8 + 30 + 9 = 20 + 30 + 8 + 9 = 50 + 17 = 67

67 – 39 = 67 – 30 – 9 = 37 – 9 = 28

Add or Subtract Numbers in row

	a	b	c	d
# 1	67 – 0	47 + 20	67 – 20	67 - 40
# 2	67 – 1	46 + 21	67 – 21	67 - 41
# 3	67 – 2	45 + 22	67 – 22	67 - 42
# 4	67 – 3	44 + 23	67 – 23	67 - 43
# 5	67 – 4	43 + 24	67 – 24	67 - 44
# 6	67 – 5	42 + 25	67 – 25	67 - 45
# 7	67 – 6	41 + 26	67 – 26	67 - 46
# 8	67 – 7	40 + 27	67 – 27	67 - 47
# 9	67 – 8	39 + 28	67 – 28	67 - 48
# 10	67 – 9	38 + 29	67 – 29	67 - 49
# 11	67 – 10	37 + 30	67 – 30	67 – 50
# 12	67 – 11	36 + 31	67 – 31	67 - 51
# 13	67 – 12	35 + 32	67 – 32	67 – 52
# 14	67 – 13	34 + 33	67 – 33	67 - 53
# 15	67 – 14	33 + 34	67 – 34	67 – 54
# 16	67 – 15	32 + 35	67 – 35	67 - 55
# 17	67 – 16	31 + 36	67 – 36	67 – 56
# 18	67 – 17	30 + 37	67 – 37	67 – 57
# 19	67 – 18	29 + 38	67 – 38	67 - 58
# 20	67 – 19	28 + 39	67 – 39	67 – 59

Grade 3 Chapter 1 Lesson 10

Number 68

39 + 29 =30 + 9 + 20 + 9 = 30 + 20 + 9 + 9 = 50 + 18 = 68

68 – 29 = 68 – 20 – 9 = 48 – 9 = 39

29 + 39 = 20 + 9 + 30 + 9 = 20 + 30 + 9 + 9 = 50 + 18 = 68

68 – 39 = 68 – 30 – 9 = 38 – 9 = 29

Add or Subtract Numbers in row

	a	b	c	d
# 1	68 – 0	48 + 20	68 – 20	68 - 40
# 2	68 – 1	47 + 21	68 – 21	68 - 41
# 3	68 – 2	46 + 22	68 – 22	68 - 42
# 4	68 – 3	45 + 23	68 – 23	68 - 43
# 5	68 – 4	44 + 24	68 – 24	68 - 44
# 6	68 – 5	43 + 25	68 – 25	68 - 45
# 7	68 – 6	42 + 26	68 – 26	68 - 46
# 8	68 – 7	41 + 27	68 – 27	68 - 47
# 9	68 – 8	40 + 28	68 – 28	68 - 48
# 10	68 – 9	39 + 29	68 – 29	68 - 49
# 11	68 – 10	38 + 30	68 – 30	68 – 50
# 12	68 – 11	37 + 31	68 – 31	68 - 51
# 13	68 – 12	36 + 32	68 – 32	68 – 52
# 14	68 – 13	35 + 33	68 – 33	68 - 53
# 15	68 – 14	34 + 34	68 – 34	68 – 54
# 16	68 – 15	33 + 35	68 – 35	68 - 55
# 17	68 – 16	32 + 36	68 – 36	68 – 56
# 18	68 – 17	31 + 37	68 – 37	68 – 57
# 19	68 – 18	30 + 38	68 – 38	68 - 58
# 20	68 – 19	29 + 39	68 – 39	68 – 59

Grade 3 Chapter 1 Lesson 10
Number 69

40 + 29 =40 + 20 + 9 = 60 + 9 = 69

69 – 29 = 69 – 20 – 9 = 49 – 9 = 40

30 + 39 = 30 + 30 + 9 = 60 + 9 = 69

69 – 39 = 69 – 30 – 9 = 39 – 9 = 30

Add or Subtract Numbers in row

	a	b	c	d
# 1	69 – 0	49 + 20	69 – 20	69 - 40
# 2	69 – 1	48 + 21	69 – 21	69 - 41
# 3	69 – 2	47 + 22	69 – 22	69 - 42
# 4	69 – 3	46 + 23	69 – 23	69 - 43
# 5	69 – 4	45 + 24	69 – 24	69 - 44
# 6	69 – 5	44 + 25	69 – 25	69 - 45
# 7	69 – 6	43 + 26	69 – 26	69 - 46
# 8	69 – 7	42 + 27	69 – 27	69 - 47
# 9	69 – 8	41 + 28	69 – 28	69 - 48
# 10	69 – 9	40 + 29	69 – 29	69 - 49
# 11	69 – 10	39 + 30	69 – 30	69 – 50
# 12	69 – 11	38 + 31	69 – 31	69 - 51
# 13	69 – 12	37 + 32	69 – 32	69 – 52
# 14	69 – 13	36 + 33	69 – 33	69 - 53
# 15	69 – 14	35 + 34	69 – 34	69 – 54
# 16	69 – 15	34 + 35	69 – 35	69 - 55
# 17	69 – 16	33 + 36	69 – 36	69 – 56
# 18	69 – 17	32 + 37	69 – 37	69 – 57
# 19	69 – 18	31 + 38	69 – 38	69 - 58
# 20	69 – 19	30 + 39	69 – 39	69 – 59

Grade 3 Chapter 1 Lesson 11
Test 1A
Add or Subtract Numbers in row

1) 40 + 29
2) 69 – 29

3) 68 – 29
4) 39 + 29

5) 38 + 29
6) 67 – 29

7) 66 – 29
8) 37 + 29

9) 36 + 29
10) 65 – 29

11) 64 – 29
12) 35 + 29

13) 34 + 29
14) 63 – 29

15) 62 – 29
16) 33 + 29

17) 32 + 29
18) 61 – 29

19) 60 – 29
20) 31 + 29

Grade 3 Chapter 1 Lesson 11
Test 1B
Add or Subtract Numbers in row

1) 30 + 39
2) 69 – 39

3) 68 – 39
4) 29 + 39

5) 28 + 39
6) 67 – 39

7) 66 – 39
8) 27 + 39

9) 26 + 39
10) 65 – 39

11) 64 – 39
12) 25 + 39

13) 24 + 39
14) 63 – 39

15) 62 – 39
16) 23 + 39

17) 22 + 39
18) 61 – 39

19) 60 – 38
20) 22 + 38

Grade 3 Answers Chapter 1

Lesson 6 Number 60

	a, b, c, d		a, b, c, d
# 1	60, 60, 40, 20	# 11	50, 60, 30, 10
# 2	59, 60, 39, 19	# 12	49, 60, 29, 9
# 3	58, 60, 38, 18	# 13	48, 60, 28, 8
# 4	57, 60, 37, 17	# 14	47, 60, 27, 7
# 5	56, 60, 36, 16	# 15	46, 60, 26, 6
# 6	55, 60, 35, 15	# 16	45, 60, 25, 5
# 7	54, 60, 34, 14	# 17	44, 60, 24, 4
# 8	53, 60, 33, 13	# 18	43, 60, 23, 3
# 9	52, 60, 32, 12	# 19	42, 60, 22, 2
# 10	51, 60, 31, 11	# 20	41, 60, 21, 1

Lesson 6 Number 61

	a, b, c, d		a, b, c, d
# 1	61, 61, 41, 21	# 11	51, 61, 31, 11
# 2	60, 61, 40, 20	# 12	50, 61, 30, 10
# 3	59, 61, 39, 19	# 13	49, 61, 29, 9
# 4	58, 61, 38, 18	# 14	48, 61, 28, 8
# 5	57, 61, 37, 17	# 15	47, 61, 27, 7
# 6	56, 61, 36, 16	# 16	46, 61, 26, 6
# 7	55, 61, 35, 15	# 17	45, 61, 25, 5
# 8	54, 61, 34, 14	# 18	44, 61, 24, 4
# 9	53, 61, 33, 13	# 19	43, 61, 23, 3
# 10	52, 61, 32, 12	# 20	42, 61, 22, 2

Lesson 7 Number 62

	a, b, c, d		a, b, c, d
# 1	62, 62, 42, 22	# 11	52, 62, 32, 12
# 2	61, 62, 41, 21	# 12	51, 62, 31, 11
# 3	60, 62, 40, 20	# 13	50, 62, 30, 10
# 4	59, 62, 39, 19	# 14	49, 62, 29, 9
# 5	58, 62, 38, 18	# 15	48, 62, 28, 8
# 6	57, 62, 37, 17	# 16	47, 62, 27, 7
# 7	56, 62, 36, 16	# 17	46, 62, 26, 6
# 8	55, 62, 35, 15	# 18	45, 62, 25, 5
# 9	54, 62, 34, 14	# 19	44, 62, 24, 4
# 10	53, 62, 33, 13	# 20	43, 62, 23, 3

Lesson 7 Number 63

	a b c d		a b c d
# 1	63, 63, 43, 23	# 11	53, 63, 33, 13
# 2	62, 63, 42, 22	# 12	52, 63, 32, 12
# 3	61, 63, 41, 21	# 13	51, 63, 31, 11
# 4	60, 63, 40, 20	# 14	50, 63, 30, 10
# 5	59, 63, 39, 19	# 15	49, 63, 29, 9
# 6	58, 63, 38, 18	# 16	48, 63, 28, 8
# 7	57, 63, 37, 17	# 17	47, 63, 27, 7
# 8	56, 63, 36, 16	# 18	46, 63, 26, 6
# 9	55, 63, 35, 15	# 19	45, 63, 25, 5
# 10	54, 63, 34, 14	# 20	44, 63, 24, 4

Lesson 8 Number 64

	a b c d		a b c d
# 1	64, 64, 44, 24	# 11	54, 64, 34, 14
# 2	63, 64, 43, 23	# 12	53, 64, 33, 13
# 3	62, 64, 42, 22	# 13	52, 64, 32, 12
# 4	61, 64, 41, 21	# 14	51, 64, 31, 11
# 5	60, 64, 40, 20	# 15	50, 64, 30, 10
# 6	59, 64, 39, 19	# 16	49, 64, 29, 9
# 7	58, 64, 38, 18	# 17	48, 64, 28, 8
# 8	57, 64, 37, 17	# 18	47, 64, 27, 7
# 9	56, 64, 36, 16	# 19	46, 64, 26, 6
# 10	55, 64, 35, 15	# 20	45, 64, 25, 5

Lesson 8 Number 65

	a b c d		a b c d
# 1	65, 65, 45, 25	# 11	55, 65, 35, 15
# 2	64, 65, 44, 24	# 12	54, 65, 34, 14
# 3	63, 65, 43, 23	# 13	53, 65, 33, 13
# 4	62, 65, 42, 22	# 14	52, 65, 32, 12
# 5	61, 65, 41, 21	# 15	51, 65, 31, 11
# 6	60, 65, 40, 20	# 16	50, 65, 30, 10
# 7	59, 65, 39, 19	# 17	49, 65, 29, 9
# 8	58, 65, 38, 18	# 18	48, 65, 28, 8
# 9	57, 65, 37, 17	# 19	47, 65, 27, 7
# 10	56, 65, 36, 16	# 20	46, 65, 26, 6

Grade 3 Answers Chapter 1

Lesson 9 Number 66

	a b c d		a b c d
# 1	66, 66, 46, 26	# 11	56, 66, 36, 16
# 2	65, 66, 45, 25	# 12	55, 66, 35, 15
# 3	64, 66, 44, 24	# 13	54, 66, 34, 14
# 4	63, 66, 43, 23	# 14	53, 66, 33, 13
# 5	62, 66, 42, 22	# 15	52, 66, 32, 12
# 6	61, 66, 41, 21	# 16	51, 66, 31, 11
# 7	60, 66, 40, 20	# 17	50, 66, 30, 10
# 8	59, 66, 39, 19	# 18	49, 66, 29, 9
# 9	58, 66, 38, 18	# 19	48, 66, 28, 8
# 10	57, 66, 37, 17	# 20	47, 66, 27, 7

Lesson 9 Number 67

	a b c d		a b c d
# 1	67, 67, 47, 27	# 11	57, 67, 37, 17
# 2	66, 67, 46, 26	# 12	56, 67, 36, 16
# 3	65, 67, 45, 25	# 13	55, 67, 35, 15
# 4	64, 67, 44, 24	# 14	54, 67, 34, 14
# 5	63, 67, 43, 23	# 15	53, 67, 33, 13
# 6	62, 67, 42, 22	# 16	52, 67, 32, 12
# 7	61, 67, 41, 21	# 17	51, 67, 31, 11
# 8	60, 67, 40, 20	# 18	50, 67, 30, 10
# 9	59, 67, 39, 19	# 19	49, 67, 29, 9
# 10	58, 67, 38, 18	# 20	48, 67, 28, 8

Lesson 10 Number 68

	a b c d		a b c d
# 1	68, 68, 48, 28	# 11	58, 68, 38, 18
# 2	67, 68, 47, 27	# 12	57, 68, 37, 17
# 3	66, 68, 46, 26	# 13	56, 68, 36, 16
# 4	65, 68, 45, 25	# 14	55, 68, 35, 15
# 5	64, 68, 44, 24	# 15	54, 68, 34, 14
# 6	63, 68, 43, 23	# 16	53, 68, 33, 13
# 7	62, 68, 42, 22	# 17	52, 68, 32, 12
# 8	61, 68, 41, 21	# 18	51, 68, 31, 11
# 9	60, 68, 40, 20	# 19	50, 68, 30, 10
# 10	59, 68, 39, 19	# 20	49, 68, 29, 9

Lesson 10 Number 69

	a b c d		a b c d
# 1	69, 69, 49, 29	# 11	59, 69, 39, 19
# 2	68, 69, 48, 28	# 12	58, 69, 38, 18
# 3	67, 69, 47, 27	# 13	57, 69, 37, 17
# 4	66, 69, 46, 26	# 14	56, 69, 36, 16
# 5	65, 69, 45, 25	# 15	55, 69, 35, 15
# 6	64, 69, 44, 24	# 16	54, 69, 34, 14
# 7	63, 69, 43, 23	# 17	53, 69, 33, 13
# 8	62, 69, 42, 22	# 18	52, 69, 32, 12
# 9	61, 69, 41, 21	# 19	51, 69, 31, 11
# 10	60, 69, 40, 20	# 20	50, 69, 30, 10

Lesson 11 Test 1A Add or Subtract Numbers in Row

1)	69	11)	35
2)	40	12)	64
3)	39	13)	63
4)	68	14)	34
5)	67	15)	33
6)	38	16)	62
7)	37	17)	61
8)	66	18)	32
9)	65	19)	31
10)	36	20)	60

Lesson 11 Test 1B Add or Subtract Numbers in Row

1)	69	11)	25
2)	30	12)	64
3)	29	13)	63
4)	68	14)	24
5)	67	15)	23
6)	28	16)	62
7)	27	17)	61
8)	66	18)	22
9)	65	19)	22
10)	26	20)	60

Grade 3

Chapter 2 Addition or Subtraction Numbers (90-99)

Grade 3 Chapter 2 Contents
Numbers 90 – 99
Lesson 1 - 11
Basic Problems

Work with lesson 1, read rules and examples, 2) Solve the problems from contents, 3) Check answers in the end of chapter 1, 4) Work with the next lessons 2, 3, 4, 5, …, 5) Solve Test 1 A, 6) Grading Scale: A =10 points, B=9points, C=8 points, D=7 points, F=0-6 points, 7) If you solve 10 problems correctly, you get an A, 8) If have an A, you can move on to next chapter, 9) If your grade is lower than an A, work with the mistakes, then solve Test 1B, 10) When you finish all of the chapters, solve Final Exam in last chapter.

Grade 3 Chapter 2 Lesson 1
Number 90

9 tens – 1 ten = 8 tens 1 ten = 10 ones

1	8 10
56	90
+34	- 34
90	56

Add or Subtract Numbers in column

	a	b	c	d	e
# 1	80 +10	90 -10	60 + 30	90 - 30	90 - 50
# 2	79 +11	90 - 11	59 +31	90 - 31	90 - 51
# 3	78 +12	90 -12	58 + 32	90 - 32	90 - 52
# 4	77 +13	90 -13	57 + 33	90 - 33	90 - 53
# 5	76 +14	90 - 14	56 + 34	90 - 34	90 - 54
# 6	75 +15	90 - 15	55 + 35	90 - 35	90 - 55
# 7	74 + 16	90 - 16	54 + 36	90 - 36	90 - 56
# 8	73 + 17	90 - 17	53 + 37	90 - 37	90 - 57
# 9	72 + 18	90 - 18	52 + 38	90 - 38	90 - 58
# 10	71 + 19	90 - 19	51 + 39	90 - 39	90 - 59

Grade 3 Chapter 2 Lesson 1
Number 90

9 tens – 1 ten = 8 tens   1 ten = 10 ones

1	8 10
46	90
+44	- 44
90	46

Add or Subtract Numbers in column

	a	b	c	d	e
# 11	70 +20	90 -20	50 + 40	90 - 40	90 - 60
# 12	69 +21	90 - 21	49 + 41	90 - 41	90 - 61
# 13	68 +22	90 - 22	48 + 42	90 - 42	90 - 62
# 14	67 + 23	90 - 23	47 + 43	90 - 43	90 - 63
# 15	66 + 24	90 - 24	46 + 44	90 - 44	90 - 64
# 16	65 +25	90 - 25	45 + 45	90 - 45	90 - 65
# 17	64 +26	90 - 26	44 + 46	90 - 46	90 - 66
# 18	63 +27	90 - 27	43 + 47	90 - 47	90 - 67
# 19	62 + 28	90 - 28	42 +48	90 - 48	90 - 68
# 20	61 + 29	90 - 29	41 +49	90 - 49	90 - 69

Grade 3 Chapter 2 Lesson 2
Number 91

9 tens – 1 ten = 8 tens   1 ten = 10 ones

1	8 11
57	91
+34	- 34
91	57

Add or Subtract Numbers in column

	a	b	c	d	e
# 1	81 +10	91 -10	61 + 30	91 - 30	91 - 50
# 2	80 +11	91 - 11	60 +31	91 - 31	91 - 51
# 3	79 +12	91 -12	59 + 32	91 - 32	91 - 52
# 4	78 +13	91 -13	58 + 33	91 - 33	91 - 53
# 5	77 +14	91 - 14	57 + 34	91 - 34	91 - 54
# 6	76 +15	91 - 15	56 + 35	91 - 35	91 - 55
# 7	75 + 16	91 - 16	55 + 36	91 - 36	91 - 56
# 8	74 + 17	91 - 17	54 + 37	91 - 37	91 - 57
# 9	73 + 18	91 - 18	53 + 38	91 - 38	91 - 58
# 10	72 + 19	91 - 19	52 + 39	91 - 39	91 - 59

Grade 3 Chapter 2 Lesson 2
Number 91

9 tens – 1 ten = 8 tens   1 ten = 10 ones

1	8 11
47	91
+44	- 44
91	47

Add or Subtract Numbers in column

	a	b	c	d	e
# 11	71 +20	91 -20	51 + 40	91 - 40	91 - 60
# 12	70 +21	91 - 21	50 + 41	91 - 41	91 - 61
# 13	69 +22	91 - 22	49 + 42	91 - 42	91 - 62
# 14	68 + 23	91 - 23	48 + 43	91 - 43	91 - 63
# 15	67 + 24	91 - 24	47 + 44	91 - 44	91 - 64
# 16	66 +25	91 - 25	46 + 45	91 - 45	91 - 65
# 17	65 +26	91 - 26	45 + 46	91 - 46	91 - 66
# 18	64 +27	91 - 27	44 + 47	91 - 47	91 - 67
# 19	63 + 28	91 - 28	43 +48	91 - 48	91 - 68
# 20	62 + 29	91 - 29	42 +49	91 - 49	91 - 69

Grade 3 Chapter 2 Lesson 3

Number 92

9 tens – 1 ten = 8 tens 1 ten = 10 ones

1	8 12
58	92
+34	- 34
92	58

Add or Subtract Numbers in column

	a	b	c	d	e
# 1	82 +10	92 -10	62 + 30	92 - 30	92 - 50
# 2	81 +11	92 - 11	61 +31	92 - 31	92 - 51
# 3	80 +12	92 -12	60 + 32	92 - 32	92 - 52
# 4	79 +13	92 -13	59 + 33	92 - 33	92 - 53
# 5	78 +14	92 - 14	58 + 34	92 - 34	92 - 54
# 6	77 +15	92 - 15	57 + 35	92 - 35	92 - 55
# 7	76 + 16	92 - 16	56 + 36	92 - 36	92 - 56
# 8	75 + 17	92 - 17	55 + 37	92 - 37	92 - 57
# 9	74 + 18	92 - 18	54 + 38	92 - 38	92 - 58
# 10	73 + 19	92 - 19	53 + 39	92 - 39	92 - 59

Grade 3 Chapter 2 Lesson 3
Number 92

9 tens – 1 ten = 8 tens 1 ten = 10 ones

1	8 12
48	92
+44	- 44
92	48

Add or Subtract Numbers in column

	a	b	c	d	e
# 11	72 +20	92 -20	52 + 40	92 - 40	92 - 60
# 12	71 +21	92 - 21	51 + 41	92 - 41	92 - 61
# 13	70 +22	92 - 22	50 + 42	92 - 42	92 - 62
# 14	69 + 23	92 - 23	49 + 43	92 - 43	92 - 63
# 15	68 + 24	92 - 24	48 + 44	92 - 44	92 - 64
# 16	67 +25	92 - 25	47 + 45	92 - 45	92 - 65
# 17	66 +26	92 - 26	46 + 46	92 - 46	92 - 66
# 18	65 +27	92 - 27	45 + 47	92 - 47	92 - 67
# 19	64 + 28	92 - 28	44 +48	92 - 48	92 - 68
# 20	63 + 29	92 - 29	43 +49	92 - 49	92 - 69

Grade 3 Chapter 2 Lesson 4
Number 93

9 tens – 1 ten = 8 tens   1 ten = 10 ones

1	8 13
59	93
+34	- 34
93	59

Add or Subtract Numbers in column

	a	b	c	d	e
# 1	83 +10	93 -10	63 + 30	93 - 30	93 - 50
# 2	82 +11	93 - 11	62 +31	93 - 31	93 - 51
# 3	81 +12	93 -12	61 + 32	93 - 32	93 - 52
# 4	80 +13	93 -13	60 + 33	93 - 33	93 - 53
# 5	79 +14	93 - 14	59 + 34	93 - 34	93 - 54
# 6	78 +15	93 - 15	58 + 35	93 - 35	93 - 55
# 7	77 + 16	93 - 16	57 + 36	93 - 36	93 - 56
# 8	76 + 17	93 - 17	56 + 37	93 - 37	93 - 57
# 9	75 + 18	93 - 18	55 + 38	93 - 38	93 - 58
# 10	74 + 19	93 - 19	54 + 39	93 - 39	93 - 59

Grade 3 Chapter 2 Lesson 4
Number 93

9 tens – 1 ten = 8 tens   1 ten = 10 ones

1	8 13
49	93
+44	- 44
93	49

Add or Subtract Numbers

	a	b	c	d	e
# 11	73 +20	93 -20	53 + 40	93 - 40	93 - 60
# 12	72 +21	93 - 21	52 + 41	93 - 41	93 - 61
# 13	71 +22	93 - 22	51 + 42	93 - 42	93 - 62
# 14	70 + 23	93 - 23	50 + 43	93 - 43	93 - 63
# 15	69 + 24	93 - 24	49 + 44	93 - 44	93 - 64
# 16	68 +25	93 - 25	48 + 45	93 - 45	93 - 65
# 17	67 +26	93 - 26	47 + 46	93 - 46	93 - 66
# 18	66 +27	93 - 27	46 + 47	93 - 47	93 - 67
# 19	65 + 28	93 - 28	45 +48	93 - 48	93 - 68
# 20	64 + 29	93 - 29	44 +49	93 - 49	93 - 69

Grade 3 Chapter 2 Lesson 5
Number 94

9 tens – 1 ten = 8 tens   1 ten = 10 ones

1	8 14
55	94
+39	- 39
94	55

Add or Subtract Numbers in column

	a	b	c	d	e
# 1	84 +10	94 -10	64 + 30	94 - 30	94 - 50
# 2	83 +11	94 - 11	63 +31	94 - 31	94 - 51
# 3	82 +12	94 -12	62 + 32	94 - 32	94 - 52
# 4	81 +13	94 -13	61 + 33	94 - 33	94 - 53
# 5	80 +14	94 - 14	60 + 34	94 - 34	94 - 54
# 6	79 +15	94 - 15	59 + 35	94 - 35	94 - 55
# 7	78 + 16	94 - 16	58 + 36	94 - 36	94 - 56
# 8	77 + 17	94 - 17	57 + 37	94 - 37	94 - 57
# 9	76 + 18	94 - 18	56 + 38	94 - 38	94 - 58
# 10	75 + 19	94 - 19	55 + 39	94 - 39	94 - 59

Grade 3 Chapter 2 Lesson 5

Number 94

9 tens – 1 ten = 8 tens   1 ten = 10 ones

1	8 14
45	94
+49	- 49
94	45

Add or Subtract Numbers in column

	a	b	c	d	e
# 11	74 +20	94 -20	54 + 40	94 - 40	94 - 60
# 12	73 +21	94 - 21	53 + 41	94 - 41	94 - 61
# 13	72 +22	94 - 22	52 + 42	94 - 42	94 - 62
# 14	71 + 23	94 - 23	51 + 43	94 - 43	94 - 63
# 15	70 + 24	94 - 24	50 + 44	94 - 44	94 - 64
# 16	69 +25	94 - 25	49 + 45	94 - 45	94 - 65
# 17	68 +26	94 - 26	48 + 46	94 - 46	94 - 66
# 18	67 +27	94 - 27	47 + 47	94 - 47	94 - 67
# 19	66 + 28	94 - 28	46 +48	94 - 48	94 - 68
# 20	65 + 29	94 - 29	45 +49	94 - 49	94 - 69

Grade 3 Chapter 2 Lesson 6
Number 95

9 tens – 1 ten = 8 tens   1 ten = 10 ones

1	8 15
56	95
+39	- 39
95	56

Add or Subtract Numbers in column

	a	b	c	d	e
# 1	85 +10	95 -10	65 + 30	95 - 30	95 - 50
# 2	84 +11	95 - 11	64 +31	95 - 31	95 - 51
# 3	83 +12	95 -12	63 + 32	95 - 32	95 - 52
# 4	82 +13	95 -13	62 + 33	95 - 33	95 - 53
# 5	81 +14	95 - 14	61 + 34	95 - 34	95 - 54
# 6	80 +15	95 - 15	60 + 35	95 - 35	95 - 55
# 7	79 + 16	95 - 16	59 + 36	95 - 36	95 - 56
# 8	78 + 17	95 - 17	58 + 37	95 - 37	95 - 57
# 9	77 + 18	95 - 18	57 + 38	95 - 38	95 - 58
# 10	76 + 19	95 - 19	56 + 39	95 - 39	95 - 59

Grade 3 Chapter 2 Lesson 6
Number 95

9 tens – 1 ten = 8 tens   1 ten = 10 ones

1	8 15
57	95
+38	- 38
95	57

Add or Subtract Numbers in column

	a	b	c	d	e
# 11	75 +20	95 -20	55 + 40	95 - 40	95 - 60
# 12	74 +21	95 - 21	54 + 41	95 - 41	95 - 61
# 13	73 +22	95 - 22	53 + 42	95 - 42	95 - 62
# 14	72 + 23	95 - 23	52 + 43	95 - 43	95 - 63
# 15	71 + 24	95 - 24	51 + 44	95 - 44	95 - 64
# 16	70 +25	95 - 25	50 + 45	95 - 45	95 - 65
# 17	69 +26	95 - 26	49 + 46	95 - 46	95 - 66
# 18	68 +27	95 - 27	48 + 47	95 - 47	95 - 67
# 19	67 + 28	95 - 28	47 +48	95 - 48	95 - 68
# 20	66 + 29	95 - 29	46 +49	95 - 49	95 - 69

Grade 3 Chapter 2 Lesson 7
Number 96

9 tens – 1 ten = 8 tens   1 ten = 10 ones

1	8 16
57	96
+39	- 39
96	57

Add or Subtract Numbers in column

	a	b	c	d	e
# 1	86 +10	96 -10	66 + 30	96 - 30	96 - 50
# 2	85 +11	96 - 11	65 +31	96 - 31	96 - 51
# 3	84 +12	96 -12	64 + 32	96 - 32	96 - 52
# 4	83 +13	96 -13	63 + 33	96 - 33	96 - 53
# 5	82 +14	96 - 14	62 + 34	96 - 34	96 - 54
# 6	81 +15	96 - 15	61 + 35	96 - 35	96 - 55
# 7	80 + 16	96 - 16	60 + 36	96 - 36	96 - 56
# 8	79 + 17	96 - 17	59 + 37	96 - 37	96 - 57
# 9	78 + 18	96 - 18	58 + 38	96 - 38	96 - 58
# 10	77 + 19	96 - 19	57 + 39	96 - 39	96 - 59

Grade 3 Chapter 2 Lesson 7
Number 96

9 tens – 1 ten = 8 tens   1 ten = 10 ones

1	8 16
47	96
+49	- 49
96	47

Add or Subtract Numbers in column

	a	b	c	d	e
# 11	76 +20	96 -20	56 + 40	96 - 40	96 - 60
# 12	75 +21	96 - 21	55 + 41	96 - 41	96 - 61
# 13	74 +22	96 - 22	54 + 42	96 - 42	96 - 62
# 14	73 + 23	96 - 23	53 + 43	96 - 43	96 - 63
# 15	72 + 24	96 - 24	52 + 44	96 - 44	96 - 64
# 16	71 +25	96 - 25	51 + 45	96 - 45	96 - 65
# 17	70 +26	96 - 26	50 + 46	96 - 46	96 - 66
# 18	69 +27	96 - 27	49 + 47	96 - 47	96 - 67
# 19	68 + 28	96 - 28	48 +48	96 - 48	96 - 68
# 20	67 + 29	96 - 29	47 +49	96 - 49	96 - 69

Grade 3 Chapter 2 Lesson 8
Number 97

9 tens – 1 ten = 8 tens   1 ten = 10 ones

1	817
48	97
+49	- 49
97	48

Add or Subtract Numbers in Column

	a	b	c	d	e
# 1	87 +10	97 -10	67 + 30	97 - 30	97 - 50
# 2	86 +11	97 - 11	66 +31	97 - 31	97 - 51
# 3	85 +12	97 -12	65 + 32	97 - 32	97 - 52
# 4	84 +13	97 -13	64 + 33	97 - 33	97 - 53
# 5	83 +14	97 - 14	63 + 34	97 - 34	97 - 54
# 6	82 +15	97 - 15	62 + 35	97 - 35	97 - 55
# 7	81 + 16	97 - 16	61 + 36	97 - 36	97 - 56
# 8	80 + 17	97 - 17	60 + 37	97 - 37	97 - 57
# 9	79 + 18	97 - 18	59 + 38	97 - 38	97 - 58
# 10	78 + 19	97 - 19	58 + 39	97 - 39	97 - 59

Grade 3 Chapter 2 Lesson 8
Number 97

9 tens – 1 ten = 8 tens 1 ten = 10 ones

1	8 17
58	97
+39	- 39
97	58

Add or Subtract Numbers in column

	a	b	c	d	e
# 11	77 +20	97 -20	57 + 40	97 - 40	97 - 60
# 12	76 +21	97 - 21	56 + 41	97 - 41	97 - 61
# 13	75 +22	97 - 22	55 + 42	97 - 42	97 - 62
# 14	74 + 23	97 - 23	54 + 43	97 - 43	97 - 63
# 15	73 + 24	97 - 24	53 + 44	97 - 44	97 - 64
# 16	72 +25	97 - 25	52 + 45	97 - 45	97 - 65
# 17	71 +26	97 - 26	51 + 46	97 - 46	97 - 66
# 18	70 +27	97 - 27	50 + 47	97 - 47	97 - 67
# 19	69 + 28	97 - 28	49 +48	97 - 48	97 - 68
# 20	68 + 29	97 - 29	48 +49	97 - 49	97 - 69

Grade 3 Chapter 2 Lesson 9
Number 98

9 tens – 1 ten = 8 tens   1 ten = 10 ones

1	8 18
59	98
+39	- 39
98	59

Add or Subtract Numbers in column

	a	b	c	d	e
# 1	88 +10	98 -10	68 + 30	98 - 30	98 - 50
# 2	87 +11	98 - 11	67 +31	98 - 31	98 - 51
# 3	86 +12	98 -12	66 + 32	98 - 32	98 - 52
# 4	85 +13	98 -13	65 + 33	98 - 33	98 - 53
# 5	84 +14	98 - 14	64 + 34	98 - 34	98 - 54
# 6	83 +15	98 - 15	63 + 35	98 - 35	98 - 55
# 7	82 + 16	98 - 16	62 + 36	98 - 36	98 - 56
# 8	81 + 17	98 - 17	61 + 37	98 - 37	98 - 57
# 9	80 + 18	98 - 18	60 + 38	98 - 38	98 - 58
# 10	79 + 19	98 - 19	59 + 39	98 - 39	98 - 59

Grade 3 Chapter 2 Lesson 9
Number 98

9 tens – 1 ten = 8 tens   1 ten = 10 ones

1	8 18
49	98
+49	- 49
98	49

Add or Subtract Numbers in column

	a	b	c	d	e
# 11	78	98	58	98	98
	+20	-20	+ 40	- 40	- 60
# 12	77	98	57	98	98
	+21	- 21	+ 41	- 41	- 61
# 13	76	98	56	98	98
	+22	- 22	+ 42	- 42	- 62
# 14	75	98	55	98	98
	+ 23	- 23	+ 43	- 43	- 63
# 15	74	98	54	98	98
	+ 24	- 24	+ 44	- 44	- 64
# 16	73	98	53	98	98
	+25	- 25	+ 45	- 45	- 65
# 17	72	98	52	98	98
	+26	- 26	+ 46	- 46	- 66
# 18	71	98	51	98	98
	+27	- 27	+ 47	- 47	- 67
# 19	70	98	50	98	98
	+ 28	- 28	+48	- 48	- 68
# 20	69	98	49	98	98
	+ 29	- 29	+49	- 49	- 69

Grade 3 Chapter 2 Lesson 10
Number 99

65	99
+34	- 34
99	65

Add or Subtract Numbers in column

	a	b	c	d	e
# 1	89 +10	99 -10	69 + 30	99 - 30	99 - 50
# 2	88 +11	99 - 11	68 +31	99 - 31	99 - 51
# 3	87 +12	99 -12	67 + 32	99 - 32	99 - 52
# 4	86 +13	99 -13	66 + 33	99 - 33	99 - 53
# 5	85 +14	99 - 14	65 + 34	99 - 34	99 - 54
# 6	84 +15	99 - 15	64 + 35	99 - 35	99 - 55
# 7	83 + 16	99 - 16	63 + 36	99 - 36	99 - 56
# 8	82 + 17	99 - 17	62 + 37	99 - 37	99 - 57
# 9	81 + 18	99 - 18	61 + 38	99 - 38	99 - 58
# 10	80 + 19	99 - 19	60 + 39	99 - 39	99 - 59

Grade 3 Chapter 2 Lesson 10
Number 99

55	99
+44	- 44
99	55

Add or Subtract Numbers in column

	a	b	c	d	e
# 11	79 +20	99 -20	59 + 40	99 - 40	99 - 60
# 12	78 +21	99 - 21	58 + 41	99 - 41	99 - 61
# 13	77 +22	99 - 22	57 + 42	99 - 42	99 - 62
# 14	76 + 23	99 - 23	56 + 43	99 - 43	99 - 63
# 15	75 + 24	99 - 24	55 + 44	99 - 44	99 - 64
# 16	74 +25	99 - 25	54 + 45	99 - 45	99 - 65
# 17	73 +26	99 - 26	53 + 46	99 - 46	99 - 66
# 18	72 +27	99 - 27	52 + 47	99 - 47	99 - 67
# 19	71 + 28	99 - 28	51 +48	99 - 48	99 - 68
# 20	70 + 29	99 - 29	50 +49	99 - 49	99 - 69

Grade 3 Chapter 2 Lesson 11
Test 2 A
Add or Subtract Numbers in Column

1) a)	56 +34	b)	90 - 34
2) a)	46 +44	b)	90 - 44
3) a)	57 +34	b)	91 - 34
4) a)	47 +44	b)	91 - 44
5) a)	58 +34	b)	92 - 34
6) a)	48 +44	b)	92 - 44
7) a)	59 +34	b)	93 - 34
8) a)	49 +44	b)	93 - 44
9) a)	55 +39	b)	94 - 39
10) a)	45 +49	b)	94 - 49

Grade 3 Chapter 2 Lesson 11
Test 2 B
Add or Subtract Numbers in Column

1) a) $\begin{array}{r} 56 \\ +39 \\ \hline \end{array}$ b) $\begin{array}{r} 95 \\ -39 \\ \hline \end{array}$

2) a) $\begin{array}{r} 57 \\ +38 \\ \hline \end{array}$ b) $\begin{array}{r} 95 \\ -38 \\ \hline \end{array}$

3) a) $\begin{array}{r} 57 \\ +39 \\ \hline \end{array}$ b) $\begin{array}{r} 96 \\ -39 \\ \hline \end{array}$

4) a) $\begin{array}{r} 47 \\ +49 \\ \hline \end{array}$ b) $\begin{array}{r} 96 \\ -49 \\ \hline \end{array}$

5) a) $\begin{array}{r} 48 \\ +49 \\ \hline \end{array}$ b) $\begin{array}{r} 97 \\ -49 \\ \hline \end{array}$

6) a) $\begin{array}{r} 58 \\ +39 \\ \hline \end{array}$ b) $\begin{array}{r} 97 \\ -39 \\ \hline \end{array}$

7) a) $\begin{array}{r} 59 \\ +39 \\ \hline \end{array}$ b) $\begin{array}{r} 98 \\ -39 \\ \hline \end{array}$

8) a) $\begin{array}{r} 49 \\ +49 \\ \hline \end{array}$ b) $\begin{array}{r} 98 \\ -49 \\ \hline \end{array}$

9) a) $\begin{array}{r} 65 \\ +34 \\ \hline \end{array}$ b) $\begin{array}{r} 99 \\ -34 \\ \hline \end{array}$

10) a) $\begin{array}{r} 60 \\ +39 \\ \hline \end{array}$ b) $\begin{array}{r} 99 \\ -39 \\ \hline \end{array}$

Grade 3 Answers Chapter 2

Lesson 1 Number 90

	a b c d e		a b c d e
# 1	90, 80, 90, 60, 40	# 11	90, 70, 90, 50, 30
# 2	90, 79, 90, 59, 39	# 12	90, 69, 90, 49, 29
# 3	90, 78, 90, 58, 38	# 13	90, 68, 90, 48, 28
# 4	90, 77, 90, 57, 37	# 14	90, 67, 90, 47, 27
# 5	90, 76, 90, 56, 36	# 15	90, 66, 90, 46, 26
# 6	90, 75, 90, 55, 35	# 16	90, 65, 90, 45, 25
# 7	90, 74, 90, 54, 34	# 17	90, 64, 90, 44, 24
# 8	90, 73, 90, 53, 33	# 18	90, 63, 90, 43, 23
# 9	90, 72, 90, 52, 32	# 19	90, 62, 90, 42, 22
# 10	90, 71, 90, 51, 31	# 20	90, 61, 90, 41, 21

Lesson 2 Number 91

	a b c d e		a b c d e
# 1	91, 81, 91, 61, 41	# 11	91, 71, 91, 51, 31
# 2	91, 80, 91, 60, 40	# 12	91, 70, 91, 50, 30
# 3	91, 79, 91, 59, 39	# 13	91, 69, 91, 49, 29
# 4	91, 78, 91, 58, 38	# 14	91, 68, 91, 48, 28
# 5	91, 77, 91, 57, 37	# 15	91, 67, 91, 47, 27
# 6	91, 76, 91, 56, 36	# 16	91, 66, 91, 46, 26
# 7	91, 75, 91, 55, 35	# 17	91, 65, 91, 45, 25
# 8	91, 74, 91, 54, 34	# 18	91, 64, 91, 44, 24
# 9	91, 73, 91, 53, 33	# 19	91, 63, 91, 43, 23
# 10	91, 72, 91, 52, 32	# 20	91, 62, 91, 42, 22

Lesson 3 Number 92

	a b c d e		a b c d e
# 1	92, 82, 92, 62, 42	# 11	92, 72, 92, 52, 32
# 2	92, 81, 92, 61, 41	# 12	92, 71, 92, 51, 31
# 3	92, 80, 92, 60, 40	# 13	92, 70, 92, 50, 30
# 4	92, 79, 92, 59, 39	# 14	92, 69, 92, 49, 29
# 5	92, 78, 92, 58, 38	# 15	92, 68, 92, 48, 28
# 6	92, 77, 92, 57, 37	# 16	92, 67, 92, 47, 27
# 7	92, 76, 92, 56, 36	# 17	92, 66, 92, 46, 26
# 8	92, 75, 92, 55, 35	# 18	92, 65, 92, 45, 25
# 9	92, 74, 92, 54, 34	# 19	92, 64, 92, 44, 24
# 10	92, 73, 92, 53, 33	# 20	92, 63, 92, 43, 23

Grade 3 Answers Chapter 2

Lesson 4 Number 93

	a, b, c, d, e		a, b, c, d, e
# 1	93, 83, 93, 63, 43	# 11	93, 73, 93, 53, 33
# 2	93, 82, 93, 62, 42	# 12	93, 72, 93, 52, 32
# 3	93, 81, 93, 61, 41	# 13	93, 71, 93, 51, 31
# 4	93, 80, 93, 60, 40	# 14	93, 70, 93, 50, 30
# 5	93, 79, 93, 59, 39	# 15	93, 69, 93, 49, 29
# 6	93, 78, 93, 58, 38	# 16	93, 68, 93, 48, 28
# 7	93, 77, 93, 57, 37	# 17	93, 67, 93, 47, 27
# 8	93, 76, 93, 56, 36	# 18	93, 66, 93, 46, 26
# 9	93, 75, 93, 55, 35	# 19	93, 65, 93, 45, 25
# 10	93, 74, 93, 54, 34	# 20	93, 64, 93, 44, 24

Lesson 5 Number 94

	a, b, c, d, e		a, b, c, d, e
# 1	94, 84, 94, 64, 44	# 11	94, 74, 94, 54, 34
# 2	94, 83, 94, 63, 43	# 12	94, 73, 94, 53, 33
# 3	94, 82, 94, 62, 42	# 13	94, 72, 94, 52, 32
# 4	94, 81, 94, 61, 41	# 14	94, 71, 94, 51, 31
# 5	94, 80, 94, 60, 40	# 15	94, 70, 94, 50, 30
# 6	94, 79, 94, 59, 39	# 16	94, 69, 94, 49, 29
# 7	94, 78, 94, 58, 38	# 17	94, 68, 94, 48, 28
# 8	94, 77, 94, 57, 37	# 18	94, 67, 94, 47, 27
# 9	94, 76, 94, 56, 36	# 19	94, 66, 94, 46, 26
# 10	94, 75, 94, 55, 35	# 20	94, 65, 94, 45, 25

Lesson 6 Number 95

	a, b, c, d, e		a, b, c, d, e
# 1	95, 85, 95, 65, 45	# 11	95, 75, 95, 55, 35
# 2	95, 84, 95, 64, 44	# 12	95, 74, 95, 54, 34
# 3	95, 83, 95, 63, 43	# 13	95, 73, 95, 53, 33
# 4	95, 82, 95, 62, 42	# 14	95, 72, 95, 52, 32
# 5	95, 81, 95, 61, 41	# 15	95, 71, 95, 51, 31
# 6	95, 80, 95, 60, 40	# 16	95, 70, 95, 50, 30
# 7	95, 79, 95, 59, 39	# 17	95, 69, 95, 49, 29
# 8	95, 78, 95, 58, 38	# 18	95, 68, 95, 48, 28
# 9	95, 77, 95, 57, 37	# 19	95, 67, 95, 47, 27
# 10	95, 76, 95, 56, 36	# 20	95, 66, 95, 46, 26

Grade 3 Answers Chapter 2

Lesson 7 Number 96

	a b c d e
# 1	96, 86, 96, 66, 46
# 2	96, 85, 96, 65, 45
# 3	96, 84, 96, 64, 44
# 4	96, 83, 96, 63, 43
# 5	96, 82, 96, 62, 42
# 6	96, 81, 96, 61, 41
# 7	96, 80, 96, 60, 40
# 8	96, 79, 96, 59, 39
# 9	96, 78, 96, 58, 38
# 10	96, 77, 96, 57, 37

	a b c d e
# 11	96, 76, 96, 56, 36
# 12	96, 75, 96, 55, 35
# 13	96, 74, 96, 54, 34
# 14	96, 73, 96, 53, 33
# 15	96, 72, 96, 52, 32
# 16	96, 71, 96, 51, 31
# 17	96, 70, 96, 50, 30
# 18	96, 69, 96, 49, 29
# 19	96, 68, 96, 48, 28
# 20	96, 67, 96, 47, 27

Lesson 8 Number 97

	a b c d e
# 1	97, 87, 97, 67, 47
# 2	97, 86, 97, 66, 46
# 3	97, 85, 97, 65, 45
# 4	97, 84, 97, 64, 44
# 5	97, 83, 97, 63, 43
# 6	97, 82, 97, 62, 42
# 7	97, 81, 97, 61, 41
# 8	97, 80, 97, 60, 40
# 9	97, 79, 97, 59, 39
# 10	97, 78, 97, 58, 38

	a b c d e
# 11	97, 77, 97, 57, 37
# 12	97, 76, 97, 56, 36
# 13	97, 75, 97, 55, 35
# 14	97, 74, 97, 54, 34
# 15	97, 73, 97, 53, 33
# 16	97, 72, 97, 52, 32
# 17	97, 71, 97, 51, 31
# 18	97, 70, 97, 50, 30
# 19	97, 69, 97, 49, 29
# 20	97, 68, 97, 48, 28

Lesson 9 Number 98

	a b c d e
# 1	98, 88, 98, 68, 48
# 2	98, 87, 98, 67, 47
# 3	98, 86, 98, 66, 46
# 4	98, 85, 98, 65, 45
# 5	98, 84, 98, 64, 44
# 6	98, 83, 98, 63, 43
# 7	98, 82, 98, 62, 42
# 8	98, 81, 98, 61, 41
# 9	98, 80, 98, 60, 40
# 10	98, 79, 98, 59, 39

	a b c d e
# 11	98, 78, 98, 58, 38
# 12	98, 77, 98, 57, 37
# 13	98, 76, 98, 56, 36
# 14	98, 75, 98, 55, 35
# 15	98, 74, 98, 54, 34
# 16	98, 73, 98, 53, 33
# 17	98, 72, 98, 52, 32
# 18	98, 71, 98, 51, 31
# 19	98, 70, 98, 50, 30
# 20	98, 69, 98, 49, 29

Lesson 10 Number 99

	a	b	c	d	e
# 1	99,	89,	99,	69,	49
# 2	99,	88,	99,	68,	48
# 3	99,	87,	99,	67,	47
# 4	99,	86,	99,	66,	46
# 5	99,	85,	99,	65,	45
# 6	99,	84,	99,	64,	44
# 7	99,	83,	99,	63,	43
# 8	99,	82,	99,	62,	42
# 9	99,	81,	99,	61,	41
# 10	99,	80,	99,	60,	40
# 11	99,	79,	99,	59,	39
# 12	99,	78,	99,	58,	38
# 13	99,	77,	99,	57,	37
# 14	99,	76,	99,	56,	36
# 15	99,	75,	99,	55,	35
# 16	99,	74,	99,	54,	34
# 17	99,	73,	99,	53,	33
# 18	99,	72,	99,	52,	32
# 19	99,	71,	99,	51,	31
# 20	99,	70,	99,	50,	30

Lesson 11 Test 2 A

1)	a)	90	b)	56
2)	a)	90	b)	46
3)	a)	91	b)	57
4)	a)	91	b)	47
5)	a)	92	b)	58
6)	a)	92	b)	48
7)	a)	93	b)	59
8)	a)	93	b)	49
9)	a)	94	b)	55
10)	a)	94	b)	45

Lesson 11 Test 2 B

1)	a)	95	b)	56
2)	a)	95	b)	57
3)	a)	96	b)	57
4)	a)	96	b)	47
5)	a)	97	b)	48
6)	a)	97	b)	58
7)	a)	98	b)	59
8)	a)	98	b)	48
9)	a)	99	b)	65
10)	a)	99	b)	60

Grade 3

Chapter 3 Addition or Subtraction Numbers (70-79)

Grade 3 Chapter 3 Contents
Numbers 70 – 79
Lesson 1 - 13
Basic Problems

Work with lesson 1, read rules and examples, 2) Solve the problems from contents, 3) Check answers in the end of chapter 1, 4) Work with the next lessons 2, 3, 4, 5, …, 5) Solve Test 1 A, 6) Grading Scale: A =10 points, B=9points, C=8 points, D=7 points, F=0-6 points, 7) If you solve 10 problems correctly, you get an A, 8) If have an A, you can move on to next chapter, 9) If your grade is lower than an A, work with the mistakes, then solve Test 1B, 10) When you finish all of the chapters, solve Final Exam in last chapter.

Grade 3 Chapter 3 Lesson 1
Number 70
7 tens – 1 ten = 6 tens   1 ten = 10 ones

1	6 10
36	70
+34	- 34
70	36

Add or Subtract Numbers in column

	a	b	c	d	e
# 1	60 +10	70 -10	40 + 30	70 - 30	70 - 50
# 2	59 +11	70 - 11	39 +31	70 - 31	70 - 51
# 3	58 +12	70 -12	38 + 32	70 - 32	70 - 52
# 4	57 +13	70 -13	37 + 33	70 - 33	70 - 53
# 5	56 +14	70 - 14	36 + 34	70 - 34	70 - 54
# 6	55 +15	70 - 15	35 + 35	70 - 35	70 - 55
# 7	54 + 16	70 - 16	34 + 36	70 - 36	70 - 56
# 8	53 + 17	70 - 17	33 + 37	70 - 37	70 - 57
# 9	52 + 18	70 - 18	32 + 38	70 - 38	70 - 58
# 10	51 + 19	70 - 19	31 + 39	70 - 39	70 - 59

Grade 3 Chapter 3 Lesson 2
Number 71

7 tens – 1 ten = 6 tens   1 ten = 10 ones

1	6 11
37	71
+34	- 34
71	37

Add or Subtract Numbers in column

	a	b	c	d	e
# 1	61 +10	71 -10	41 + 30	71 - 30	71 - 50
# 2	60 +11	71 - 11	40 +31	71 - 31	71 - 51
# 3	59 +12	71 -12	39 + 32	71 - 32	71 - 52
# 4	58 +13	71 -13	38 + 33	71 - 33	71 - 53
# 5	57 +14	71 - 14	37 + 34	71 - 34	71 - 54
# 6	56 +15	71 - 15	36 + 35	71 - 35	71 - 55
# 7	55 + 16	71 - 16	35 + 36	71 - 36	71 - 56
# 8	54 + 17	71 - 17	34 + 37	71 - 37	71 - 57
# 9	53 + 18	71 - 18	33 + 38	71 - 38	71 - 58
# 10	52 + 19	71 - 19	32 + 39	71 - 39	71 - 59

Grade 3 Chapter 3 Lesson 3
Number 72

7 tens – 1 ten = 6 tens   1 ten = 10 ones

1	6 12
38	72
+34	- 34
72	38

Add or Subtract Numbers in column

	a	b	c	d	e
# 1	62 +10	72 -10	42 + 30	72 - 30	72 - 50
# 2	61 +11	72 - 11	41 +31	72 - 31	72 - 51
# 3	60 +12	72 -12	40 + 32	72 - 32	72 - 52
# 4	59 +13	72 -13	39 + 33	72 - 33	72 - 53
# 5	58 +14	72 - 14	38 + 34	72 - 34	72 - 54
# 6	57 +15	72 - 15	37 + 35	72 - 35	72 - 55
# 7	56 + 16	72 - 16	36 + 36	72 - 36	72 - 56
# 8	55 + 17	72 - 17	35 + 37	72 - 37	72 - 57
# 9	54 + 18	72 - 18	34 + 38	72 - 38	72 - 58
# 10	53 + 19	72 - 19	33 + 39	72 - 39	72 - 59

Grade 3 Chapter 3 Lesson 4
Number 73

7 tens – 1 ten = 6 tens   1 ten = 10 ones

```
  1        6 13
  39        73
 +34      - 34
 ---      ----
  73        39
```

Add or Subtract Numbers in column

	a	b	c	d	e
# 1	63 +10	73 -10	43 + 30	73 - 30	73 - 50
# 2	62 +11	73 - 11	42 +31	73 - 31	73 - 51
# 3	61 +12	73 -12	41 + 32	73 - 32	73 - 52
# 4	60 +13	73 -13	40 + 33	73 - 33	73 - 53
# 5	59 +14	73 - 14	39 + 34	73 - 34	73 - 54
# 6	58 +15	73 - 15	38 + 35	73 - 35	73 - 55
# 7	57 + 16	73 - 16	37 + 36	73 - 36	73 - 56
# 8	56 + 17	73 - 17	36 + 37	73 - 37	73 - 57
# 9	55 + 18	73 - 18	35 + 38	73 - 38	73 - 58
# 10	54 + 19	73 - 19	34 + 39	73 - 39	73 - 59

Grade 3 Chapter 3 Lesson 5

Number 74

7 tens – 1 ten = 6 tens 1 ten = 10 ones

1	6 14
35	74
+39	- 39
74	35

Add or Subtract Numbers in column

	a	b	c	d	e
# 1	64	74	44	74	74
	+10	-10	+ 30	- 30	- 50
# 2	63	74	43	74	74
	+11	- 11	+31	- 31	- 51
# 3	62	74	42	74	74
	+12	-12	+ 32	- 32	- 52
# 4	61	74	41	74	74
	+13	-13	+ 33	- 33	- 53
# 5	60	74	40	74	74
	+14	- 14	+ 34	- 34	- 54
# 6	59	74	39	74	74
	+15	- 15	+ 35	- 35	- 55
# 7	58	74	38	74	74
	+16	- 16	+ 36	- 36	- 56
# 8	57	74	37	74	74
	+ 17	- 17	+ 37	- 37	- 57
# 9	56	74	36	74	74
	+ 18	- 18	+ 38	- 38	- 58
# 10	55	74	35	74	74
	+ 19	- 19	+ 39	- 39	- 59

Grade 3 Chapter 3 Lesson 6
Number 75

7 tens – 1 ten = 6 tens   1 ten = 10 ones

1	6 15
36	75
+39	- 39
75	36

Add or Subtract Numbers in column

	a	b	c	d	e
# 1	65 +10	75 -10	45 + 30	75 - 30	75 - 50
# 2	64 +11	75 - 11	44 +31	75 - 31	75 - 51
# 3	63 +12	75 -12	43 + 32	75 - 32	75 - 52
# 4	62 +13	75 -13	42 + 33	75 - 33	75 - 53
# 5	61 +14	75 - 14	41 + 34	75 - 34	75 - 54
# 6	60 +15	75 - 15	40 + 35	75 - 35	75 - 55
# 7	59 + 16	75 - 16	39 + 36	75 - 36	75 - 56
# 8	58 + 17	75 - 17	38 + 37	75 - 37	75 - 57
# 9	57 + 18	75 - 18	37 + 38	75 - 38	75 - 58
# 10	56 + 19	75 - 19	36 + 39	75 - 39	75 - 59

Grade 3 Chapter 3 Lesson 7
Number 76
7 tens – 1 ten = 6 tens   1 ten = 10 ones

1	6 16
37	76
+39	- 39
76	37

Add or Subtract Numbers in column

	a	b	c	d	e
# 1	66 +10	76 -10	46 + 30	76 - 30	76 - 50
# 2	65 +11	76 - 11	45 +31	76 - 31	76 - 51
# 3	64 +12	76 -12	44 + 32	76 - 32	76 - 52
# 4	63 +13	76 -13	43 + 33	76 - 33	76 - 53
# 5	62 +14	76 - 14	42 + 34	76 - 34	76 - 54
# 6	61 +15	76 - 15	41 + 35	76 - 35	76 - 55
# 7	60 + 16	76 - 16	40 + 36	76 - 36	76 - 56
# 8	59 + 17	76 - 17	39 + 37	76 - 37	76 - 57
# 9	58 + 18	76 - 18	38 + 38	76 - 38	76 - 58
# 10	57 + 19	76 - 19	37 + 39	76 - 39	76 - 59

Grade 3 Chapter 3 Lesson 8
Number 77

7 tens – 1 ten = 6 tens   1 ten = 10 ones

1	617
38	77
+39	- 39
77	38

Add or Subtract Numbers in Column

	a	b	c	d	e
# 1	67 +10	77 -10	47 + 30	77 - 30	77 - 50
# 2	66 +11	77 - 11	46 +31	77 - 31	77 - 51
# 3	65 +12	77 -12	45 + 32	77 - 32	77 - 52
# 4	64 +13	77 -13	44 + 33	77 - 33	77 - 53
# 5	63 +14	77 - 14	43 + 34	77 - 34	77 - 54
# 6	62 +15	77 - 15	42 + 35	77 - 35	77 - 55
# 7	61 + 16	77 - 16	41 + 36	77 - 36	77 - 56
# 8	60 + 17	77 - 17	40 + 37	77 - 37	77 - 57
# 9	59 + 18	77 - 18	39 + 38	77 - 38	77 - 58
# 10	58 + 19	77 - 19	38 + 39	77 - 39	77 - 59

Grade 3 Chapter 3 Lesson 9
Number 78

7 tens – 1 ten = 6 tens   1 ten = 10 ones

1	6 18
39	78
+39	- 39
78	39

Add or Subtract Numbers in column

	a	b	c	d	e
# 1	68 +10	78 -10	48 + 30	78 - 30	78 - 50
# 2	67 +11	78 - 11	47 +31	78 - 31	78 - 51
# 3	66 +12	78 -12	46 + 32	78 - 32	78 - 52
# 4	65 +13	78 -13	45 + 33	78 - 33	78 - 53
# 5	64 +14	78 - 14	44 + 34	78 - 34	78 - 54
# 6	63 +15	78 - 15	43 + 35	78 - 35	78 - 55
# 7	62 + 16	78 - 16	42 + 36	78 - 36	78 - 56
# 8	61 + 17	78 - 17	41 + 37	78 - 37	78 - 57
# 9	60 + 18	78 - 18	40 + 38	78 - 38	78 - 58
# 10	59 + 19	78 - 19	49 + 39	78 - 39	78 - 59

Grade 3 Chapter 3 Lesson 10
Number 79

7 tens – 1 ten = 6 tens 1 ten = 10 ones

45	79
+34	- 34
79	45

Add or Subtract Numbers in column

	a	b	c	d	e
# 1	69 +10	79 -10	49 + 30	79 - 30	79 - 50
# 2	68 +11	79 - 11	48 +31	79 - 31	79 - 51
# 3	67 +12	79 -12	47 + 32	79 - 32	79 - 52
# 4	66 +13	79 -13	46 + 33	79 - 33	79 - 53
# 5	65 +14	79 - 14	45 + 34	79 - 34	79 - 54
# 6	64 +15	79 - 15	44 + 35	79 - 35	79 - 55
# 7	63 + 16	79 - 16	43 + 36	79 - 36	79 - 56
# 8	62 + 17	79 - 17	42 + 37	79 - 37	79 - 57
# 9	61 + 18	79 - 18	41 + 38	79 - 38	79 - 58
# 10	60 + 19	79 - 19	40 + 39	79 - 39	79 - 59

Grade 3 Chapter 3 Lesson 11
Test 3 A
Add or Subtract Numbers in Column

1)	a)	36 +34	b)	70 - 35
2)	a)	26 +44	b)	70 - 34
3)	a)	37 +34	b)	71 - 34
4)	a)	27 +44	b)	71 - 33
5)	a)	38 +34	b)	72 - 33
6)	a)	28 +44	b)	72 - 47
7)	a)	39 +34	b)	73 - 47
8)	a)	29 +44	b)	73 - 46
9)	a)	35 +39	b)	74 - 46
10)	a)	25 +49	b)	74 - 45

Grade 3 Chapter 3 Lesson 11
Test 3 B
Add or Subtract Numbers in Column

	a)	b)
1)	$\begin{array}{r} 36 \\ +39 \\ \hline \end{array}$	$\begin{array}{r} 75 \\ -50 \\ \hline \end{array}$
2)	$\begin{array}{r} 37 \\ +38 \\ \hline \end{array}$	$\begin{array}{r} 75 \\ -49 \\ \hline \end{array}$
3)	$\begin{array}{r} 37 \\ +39 \\ \hline \end{array}$	$\begin{array}{r} 76 \\ -49 \\ \hline \end{array}$
4)	$\begin{array}{r} 27 \\ +49 \\ \hline \end{array}$	$\begin{array}{r} 76 \\ -48 \\ \hline \end{array}$
5)	$\begin{array}{r} 28 \\ +49 \\ \hline \end{array}$	$\begin{array}{r} 77 \\ -48 \\ \hline \end{array}$
6)	$\begin{array}{r} 38 \\ +39 \\ \hline \end{array}$	$\begin{array}{r} 77 \\ -42 \\ \hline \end{array}$
7)	$\begin{array}{r} 39 \\ +39 \\ \hline \end{array}$	$\begin{array}{r} 78 \\ -42 \\ \hline \end{array}$
8)	$\begin{array}{r} 29 \\ +49 \\ \hline \end{array}$	$\begin{array}{r} 78 \\ -41 \\ \hline \end{array}$
9)	$\begin{array}{r} 45 \\ +34 \\ \hline \end{array}$	$\begin{array}{r} 79 \\ -41 \\ \hline \end{array}$
10)	$\begin{array}{r} 40 \\ +39 \\ \hline \end{array}$	$\begin{array}{r} 79 \\ -40 \\ \hline \end{array}$

Grade 3 Answers Chapter 3

Lesson 1 Number 70

	a	b	c	d	e
# 1	70,	60,	70,	40,	20
# 2	70,	59,	70,	39,	19
# 3	70,	58,	70,	38,	18
# 4	70,	57,	70,	37,	17
# 5	70,	56,	70,	36,	16
# 6	70,	55,	70,	35,	15
# 7	70,	54,	70,	34,	14
# 8	70,	53,	70,	33,	13
# 9	70,	52,	70,	32,	12
# 10	70,	51,	70,	31,	11

Lesson 2 Number 71

	a	b	c	d	e
# 1	71,	61,	71,	41,	21
# 2	71,	60,	71,	40,	20
# 3	71,	59,	71,	39,	19
# 4	71,	58,	71,	38,	18
# 5	71,	57,	71,	37,	17
# 6	71,	56,	71,	36,	16
# 7	71,	55,	71,	35,	15
# 8	71,	54,	71,	34,	14
# 9	71,	53,	71,	33,	13
# 10	71,	52,	71,	32,	12

Lesson 3 Number 72

	a	b	c	d	e
# 1	72,	62,	72,	42,	22
# 2	72,	61,	72,	41,	21
# 3	72,	60,	72,	40,	20
# 4	72,	59,	72,	39,	19
# 5	72,	58,	72,	38,	18
# 6	72,	57,	72,	37,	17
# 7	72,	56,	72,	36,	16
# 8	72,	55,	72,	35,	15
# 9	72,	54,	72,	34,	14
# 10	72,	53,	72,	33,	13

Grade 3 Answers Chapter 3

Lesson 4 Number 73

	a, b, c, d, e
# 1	73, 63, 73, 43, 23
# 2	73, 62, 73, 42, 22
# 3	73, 61, 73, 41, 21
# 4	73, 60, 73, 40, 20
# 5	73, 59, 73, 39, 19
# 6	73, 58, 73, 38, 18
# 7	73, 57, 73, 37, 17
# 8	73, 56, 73, 36, 16
# 9	73, 55, 73, 35, 15
# 10	73, 54, 73, 34, 14

Lesson 5 Number 74

	a, b, c, d, e
# 1	74, 64, 74, 44, 24
# 2	74, 63, 74, 43, 23
# 3	74, 62, 74, 42, 22
# 4	74, 61, 74, 41, 21
# 5	74, 60, 74, 40, 20
# 6	74, 59, 74, 39, 19
# 7	74, 58, 74, 38, 18
# 8	74, 57, 74, 37, 17
# 9	74, 56, 74, 36, 16
# 10	74, 55, 74, 35, 15

Lesson 6 Number 75

	a, b, c, d, e
# 1	75, 65, 75, 45, 25
# 2	75, 64, 75, 44, 24
# 3	75, 63, 75, 43, 23
# 4	75, 62, 75, 42, 22
# 5	75, 61, 75, 41, 21
# 6	75, 60, 75, 40, 20
# 7	75, 59, 75, 39, 19
# 8	75, 58, 75, 38, 18
# 9	75, 57, 75, 37, 17
# 10	75, 56, 75, 36, 16

Grade 3 Answers Chapter 3

Lesson 7 Number 76

	a	b	c	d	e
# 1	76,	66,	76,	46,	26
# 2	76,	65,	76,	45,	25
# 3	76,	64,	76,	44,	24
# 4	76,	63,	76,	43,	23
# 5	76,	62,	76,	42,	22
# 6	76,	61,	76,	41,	21
# 7	76,	60,	76,	40,	20
# 8	76,	59,	76,	39,	19
# 9	76,	58,	76,	38,	18
# 10	76,	57,	76,	37,	17

Lesson 8 Number 77

	a	b	c	d	e
# 1	77,	67,	77,	47,	27
# 2	77,	66,	77,	46,	26
# 3	77,	65,	77,	45,	25
# 4	77,	64,	77,	44,	24
# 5	77,	63,	77,	43,	23
# 6	77,	62,	77,	42,	22
# 7	77,	61,	77,	41,	21
# 8	77,	60,	77,	40,	20
# 9	77,	59,	77,	39,	19
# 10	77,	58,	77,	38,	18

Lesson 9 Number 78

	a	b	c	d	e
# 1	78,	68,	78,	48,	28
# 2	78,	67,	78,	47,	27
# 3	78,	66,	78,	46,	26
# 4	78,	65,	78,	45,	25
# 5	78,	64,	78,	44,	24
# 6	78,	63,	78,	43,	23
# 7	78,	62,	78,	42,	22
# 8	78,	61,	78,	41,	21
# 9	78,	60,	78,	40,	20
# 10	78,	59,	78,	39,	19

Grade 3 Answers Chapter 3

Lesson 10 Number 79

	a, b, c, d, e
# 1	79, 69, 79, 49, 29
# 2	79, 68, 79, 48, 28
# 3	79, 67, 79, 47, 27
# 4	79, 66, 79, 46, 26
# 5	79, 65, 79, 45, 25
# 6	79, 64, 79, 44, 24
# 7	79, 63, 79, 43, 23
# 8	79, 62, 79, 42, 22
# 9	79, 61, 79, 41, 21
# 10	79, 60, 79, 40, 20

Lesson 11 Test 3 A

1)	a)	70	b)	35
2)	a)	70	b)	36
3)	a)	71	b)	37
4)	a)	71	b)	38
5)	a)	72	b)	39
6)	a)	72	b)	25
7)	a)	73	b)	26
8)	a)	73	b)	27
9)	a)	74	b)	28
10)	a)	74	b)	29

Lesson 11 Test 3 B

1)	a)	75	b)	25
2)	a)	75	b)	26
3)	a)	76	b)	27
4)	a)	76	b)	28
5)	a)	77	b)	29
6)	a)	77	b)	35
7)	a)	78	b)	36
8)	a)	78	b)	37
9)	a)	79	b)	38
10)	a)	79	b)	39

Grade 3

Chapter 4 Finding Unknowns (80-89)

Grade 3 Chapter 4 Contents
Numbers 80 – 89 Lesson 1 - 19
Advanced Problems

Work with lesson 1, read rules and examples, 2) Solve the problems from contents, 3) Check answers in the end of chapter 1, 4) Work with the next lessons 2, 3, 4, 5, …, 5) Solve Test 1 A, 6) Grading Scale: A =10 points, B=9points, C=8 points, D=7 points, F=0-6 points, 7) If you solve 10 problems correctly, you get an A, 8) If have an A, you can move on to next chapter, 9) If your grade is lower than an A, work with the mistakes, then solve Test 1B, 10) When you finish all of the chapters, solve Final Exam in last chapter.

Grade 3 Chapter 4 Lesson 1

Summary 1

One-digit numbers

Count forward from 1 to 9

1, 2, 3, 4, 5, 6, 7, 8, 9

Count backwards from 9 to 1

9, 8, 7, 6, 5, 4, 3, 2, 1

Compare one digit numbers (< , > , =)

5 = 5   0 < 1   5 < 8   6 > 4   9 > 8

Two-digit numbers

Digits   0, 1, 2, 3, 4, 5, 6, 7, 8, 9

One-digit numbers have one digit

This is numbers from 0 to 9

Two-digit numbers have two digits

This is numbers from 10 to 99

Digit from the left called tens

Digit from the right called ones

Three-digit numbers

Three-digit numbers content three digits

This is numbers 100, 101, 102, … 999

Least three-digit number: 100

Greatest three-digit numbers: 999

Digits for three-digit numbers from right to the left:

Ones, tens, and hundreds: 238

Comparison one and two-digit numbers

One-digit numbers always less than two-digit numbers   9 < 11

Two-digit numbers always greater than one-digit numbers   19 > 8

To compare two-digit numbers

Compare the tens and then compare the ones   12<18   18 > 12

Grade 3 Chapter 4 Lesson 1

Summary 1

Numbers 20 to 39

Add or Subtract Numbers

# 1	18 + 12	# 11	30 – 22
# 2	22 + 9	# 12	31 – 12
# 3	28 + 6	# 13	34 – 27
# 4	18 + 17	# 14	35 – 19
# 5	29 + 7	# 15	36 – 9
# 6	25 + 8	# 16	33 – 14
# 7	26 + 6	# 17	32 – 24
# 8	23 + 16	# 18	39 – 28
# 9	29 + 9	# 19	38 – 19
# 10	28 + 9	# 20	37 – 18

Numbers 60 - 79

Add or Subtract numbers

# 21	41 + 29	# 31	71 – 23
# 22	15 + 57	# 32	73 – 27
# 23	19 + 55	# 33	75 – 36
# 24	18 + 58	# 34	77 – 48
# 25	29 + 49	# 35	79 - 49
# 26	58 + 13	# 36	70 – 21
# 27	56 + 17	# 37	72 – 25
# 28	59 + 16	# 38	74 – 29
# 29	49 + 28	# 39	76 – 38
# 30	39 + 40	# 40	78 - 59

Grade 3 Chapter 4 Lesson 1

Summary 2

Addition and Subtraction Numbers

Properties of zeros

Two numbers can be added in any order with the same sum

$2 + 3 = 5 \qquad 3 + 2 = 5$

The sum of a number and zero is that number

$5 + 0 = 5 \qquad 0 + 5 = 5$

Difference any number and zero is the same number

$5 - 0 = 5$

Difference two the same number is zero

$5 - 5 = 0$

Addition

First number in addition called the addend	**2**
Second number in addition called the addend	**3**
Result of addition called the sum	**5**

Subtraction

First number in subtraction called the minuend	**5**
Second number in subtraction called the subtrahend	**2**
Result of subtraction called the difference	**3**

Grade 3 Chapter 4 Lesson 1

Summary 2

Numbers 40 and 59

Add or Subtract numbers in column

# 1 23 +17	# 2 24 + 18	# 3 25 +19	# 4 29 + 17	# 5 29 +19
# 6 12 +19	#7 26 + 17	# 8 26 + 19	# 9 28 +19	# 10 35 +14
# 11 41 -28	# 12 43 -29	# 13 45 -16	# 14 47 - 19	# 15 49 - 15
# 16 40 -15	#17 42 -26	# 18 44 - 17	#19 46 -18	# 20 48 - 29

Numbers 80 - 99

Add or Subtract numbers in column

# 21 53 + 27	# 22 56 + 26	# 23 69 +15	# 24 57 +29	#25 69 + 19
# 26 67 + 14	# 27 49 + 34	# 28 67 +18	# 29 58 +29	# 30 68 +21
# 31 81 -67	# 32 83 - 58	# 33 85 - 47	# 34 87 - 39	# 35 89 - 59
# 36 80 -65	# 37 82 - 58	# 38 84 - 47	# 39 86 - 39	# 40 88 - 29

Grade 3 Chapter 4 Lesson 2
Summary 3
Addition and Subtraction Numbers

Addition

First number in addition called the addend 2

Second number in addition called the addend 3

Result of addition called the sum 5

Addends smaller than sum

Addend + Addend = Sum $2 + 3 = 5$

First Addend = Sum – Second Addend
$? + 3 = 5$ $5 - 3 = 2$
Second Addend = Sum – First Addend
$2 + ? = 5$ $5 - 2 = 3$
Sum greater than any addends

Subtraction

First number in subtraction called the minuend 5

Second number in subtraction called the subtrahend 2

Result of subtraction called the difference 3

Minuend – Subtrahend = Difference $5 - 2 = 3$

Minuend = Subtrahend + Difference
$? - 2=3$ $3 + 2 = 5$
Subtrahend = Minuend – Difference
$5 - ? = 3$ $5 - 3 = 2$
Minuend greater than subtrahend and then difference

Grade 3 Chapter 4 Lesson 2

Summary 3
Numbers from 20 to 39

Put correct number instead question mark

# 1	14 + ? = 20	# 11	20 - ? = 6
# 2	? + 4 = 21	# 12	? – 6 = 15
# 3	19 + ? = 24	# 13	24 - ? = 16
# 4	17 + ? = 25	# 14	? – 6 = 19
# 5	? + 8 = 26	# 15	? – 9 = 17
# 6	19 + ? = 23	# 16	23 - ? = 15
# 7	? + 4 = 22	# 17	? – 5 = 17
# 8	29 - ? = 24	# 18	29 - ? = 13
# 9	19 + ? = 28	# 19	? – 10 = 28
# 10	18 + ? = 27	# 20	27 - ? = 9

Numbers 60 - 79
Put correct number instead question mark

# 21	? + 39 = 60	# 31	61 - ? = 28
# 22	? + 37 = 62	# 32	63 - ? = 26
# 23	? + 35 = 64	# 33	65 - ? = 39
# 24	? + 29 = 66	# 34	67 - ? = 39
# 25	? + 29 = 68	# 35	69 - ? = 49
# 26	23 + ? = 61	# 36	? - 31 = 29
# 27	27 + ? = 63	# 37	? - 35 = 37
# 28	26 + ? = 65	# 38	? – 29 = 35
# 29	49 + ? = 67	# 39	? - 37 = 29
# 30	29 + ? = 69	# 40	? – 39 = 29

Grade 3 Chapter 4 Lesson 2
Summary 4
Two operations with numbers

Three numbers can add in any order

11 + 14 + 19 = 25 + 19 = 44 11 + 14 + 19 = 11 + 33 = 44

11 + 14 + 19 = 11 + 19 + 14 = 30 + 14 = 44

To do two operations, work from left to right

To subtract three numbers, work from left to right

10 – 1 - 6 = 9 – 6 = 3 10 – 1 – 7 = 9 – 7 = 2

19 – 1 – 9 = 18 – 9 = 9 18 – 2 – 8 = 16 – 8 = 8

To add and subtract numbers, work from left to right

5 + 8 – 9 = 13 – 9 = 4 5 + 9 – 8 = 14 – 8 = 6

19 + 9 - 1 = 26 - 1 = 25 18 + 8 - 2 = 26 - 2 = 24

To subtract and add, work from left to right

9 – 8 + 5 = 1 + 5 = 6 9 – 7 + 6 = 2 + 6 = 8

18 - 9 + 8 = 9 + 8 = 17 19 - 8 + 9 = 11 + 9 = 20

Add or Subtract Numbers

15 + 4 = 10 + 5 + 4 = 10 + 9 = 19

39 + 29 =30 + 9 + 20 + 9 = 30 + 20 + 9 + 9 = 50 + 18 = 68

68 – 29 = 68 – 20 – 9 = 48 – 9 = 39

To check addition use operation subtraction

2 + 3 = 5 Check 5 – 2 = 3 or 5 – 3 = 2

To check subtraction, use operation addition or subtraction

5 – 2 = 3 Check 2 + 3 = 5 or 5 – 3 = 2

Grade 3 Chapter 4 Lesson 2
Summary 4

Rules	Examples	
Addend + Addend = Sum	2 + 3 = 5	
First Addend = Sum – Second Addend	? + 3 = 5	5 – 3 = 2
Second Addend = Sum – First Addend	2 + ? = 5	5 – 2 = 3
Minuend – Subtrahend = Difference	5 – 2 = 3	
Minuend = Subtrahend + Difference	? - 2 = 3	3 + 2 = 5
Subtrahend = Minuend – Difference	5 - ? = 3	5 – 3 = 2

Numbers 40 and 59
Complete the tables

	#1	#2	#3	#4	# 5	#6	#7	#8	#9	# 10
Addend		16	17		19		16	27		19
Addend	15		37	18		26		28	29	
Sum	50	52		56	58	51	53		57	59

	#11	#12	#13	#14	#15	#16	#17	#18	#19	#20
Minuend		56	54		50		57	55		51
Subtrahend	39		17	36		42		28	27	
Difference	19	38		16	33	17	28		26	29

Numbers 80 and 99
Complete the tables

	#21	#22	#23	#24	# 25	#26	#27	#28	#29	# 30
Addend		19	29		29	59		67	79	
Addend	73		66	75		33	49		20	57
Sum	91	93		97	90		94	96		98

	#31	#32	#33	#34	#35	#36	#37	#38	#39	#40
Minuend		92	91		94	96		97	98	
Subtrahend	19		29	59		57	29		40	65
Difference	71	64		34	49		66	49		34

Grade 3 Chapter 4 Lesson 3

Review 1 A

Rules	Examples	
Addend + Addend = Sum	2 + 3 = 5	
First Addend = Sum – Second Addend	? + 3 = 5	5 – 3 = 2
Second Addend = Sum – First Addend	2 + ? = 5	5 – 2 = 3
Minuend – Subtrahend = Difference	5 – 2 = 3	
Minuend = Subtrahend + Difference	? - 2 = 3	3 + 2 = 5
Subtrahend = Minuend – Difference	5 - ? = 3	5 – 3 = 2

Put right number instead question mark

	a	b	c	d
# 1	2 + ? = 30	? + 29 = 30	0 + ? = 30	17 + ? = 30
# 2	30 – 28 = ?	30 - ? = 1	30 – ? = 0	30 – 14 = ?
# 3	3 + ? = 31	? + 29 = 31	1 + ? = 31	? + 31 = 31
# 4	31 – ? = 3	31 – 29 = ?	31 – ? = 1	31 – 31 = ?
# 5	? + 28 = 32	3 + ? = 32	? + 30 = 32	1 + ? = 32
# 6	32 – ? = 4	32 – 29 = ?	32 – ? = 2	32 – 31 = ?
# 7	? + 28 = 33	4 + ? = 33	? + 30 = 33	2 + ? = 33
# 8	33 – ? = 5	33 - 29 = ?	33 – ? = 3	33 – 31 = ?
# 9	? + 28 = 34	5 + ? = 34	? + 30 = 34	3 + ? = 34
# 10	34 – ? = 6	34 – 29 = ?	34 – ? = 4	34 - 31 = ?

Grade 3 Chapter 4 Lesson 3

Review 1 B

Rules	Examples	
Addend + Addend = Sum	2 + 3 = 5	
First Addend = Sum – Second Addend	? + 3 = 5	5 – 3 = 2
Second Addend = Sum – First Addend	2 + ? = 5	5 – 2 = 3
Minuend – Subtrahend = Difference	5 – 2 = 3	
Minuend = Subtrahend + Difference	? - 2 = 3	3 + 2 = 5
Subtrahend = Minuend – Difference	5 - ? = 3	5 – 3 = 2

Put right number instead question mark

	a	b	c	d
# 11	7 + 28 = ?	6 + ? = 35	? + 30 = 35	4 + ? = 35
# 12	? – 28 = 7	35 – 29 = ?	35 – ? = 5	35 – 31 = ?
# 13	8 + 28 = ?	7 + ? = 36	? + 30 = 36	5 + ? = 36
# 14	? – 28 = 8	36 – ? = 7	36 – 30 = ?	36 – ? = 5
# 15	9 + 28 = ?	8 + ? = 37	? + 30 = 37	6 + ? = 37
# 16	? – 28 = 9	37 – ? = 8	37 – 30 = ?	37 – ? = 6
# 17	10 + 28 = ?	9 + ? = 38	? + 30 = 38	7 + ? = 38
# 18	? – 28 = 10	38 – ? = 9	38 – 30 = ?	38 – ? = 7
# 19	11 + 28 = ?	10 + ? = 39	? + 30 = 39	8 + ? = 39
# 20	? – 28 = 11	39 – ? = 10	39 – 30 = ?	39 – ? = 8

Grade 3 Chapter 4 Lesson 4
Number 80 Complete the tables

Addend + Addend = Sum — 2 + 3 = 5

First Addend = Sum – Second Addend — ? + 3 = 5 — 5 – 3 = 2

Second Addend = Sum – First Addend — 2 + ? = 5 — 5 – 2 = 3

 # 1

	a	b	c	d	e
Addend	72		40		8
Addend		24	40	56	
Sum	80	80		80	80

 # 2

Addend	71		39		7
Addend		25	41	57	
Sum	80	80		80	80

 # 3

Addend	70		38		6
Addend		26	42	58	
Sum	80	80		80	80

 # 4

Addend	69		37		5
Addend		27	43	59	
Sum	80	80		80	80

 # 5

Addend	68		36		4
Addend		28	44	60	
Sum	80	80		80	80

 # 6

a) 67 + ? = 80

	a				
Addend	67		35		3
Addend	?	29	45	61	
Sum	80	80		80	80

 # 7

b) ? + 30 = 80

		b			
Addend	66	?	34		2
Addend		30	46	62	
Sum	80	80		80	80

 # 8

	a	b	c	d	e
Addend	65	?	33	?	1
Addend	?	31	47	63	?
Sum	80	80	?	80	80

Grade 3 Chapter 4 Lesson 4
Number 80 Complete the tables

Minuend – Subtrahend = Difference 5 – 2 = 3

Minuend = Subtrahend + Difference ? - 2 = 3 3 + 2 = 5

Subtrahend = Minuend – Difference 5 - ? = 3 5 – 3 = 2

 # 9

	a	b	c	d	e
Minuend	80	80		80	80
Subtrahend		16	32		64
Difference	80		48	32	

 # 10

Minuend	80	80		80	80
Subtrahend		17	33		65
Difference	79		47	31	

 # 11

Minuend	80	80		80	80
Subtrahend		18	34		66
Difference	78		46	30	

 # 12

Minuend	80	80		80	80
Subtrahend		19	35		67
Difference	77		45	29	

 # 13

Minuend	80	80		80	80
Subtrahend		20	36		68
Difference	76		44	28	

 # 14 a) 80 - ? = 75

	a				
Minuend	80	80		80	80
Subtrahend	?	21	37		69
Difference	75		43	27	

 # 15 c) ? - 38 = 42

			c		
Minuend	80	80	?	80	80
Subtrahend		22	38		70
Difference	74		42	26	

 # 16

	a	b	c	d	e
Minuend	80	80	?	80	80
Subtrahend	?	23	39	?	71
Difference	73	?	41	25	?

Grade 3 Chapter 4 Lesson 5
Number 81 Complete the tables

Addend + Addend = Sum — 2 + 3 = 5

First Addend = Sum – Second Addend — ? + 3 = 5 — 5 - 3 = 2

Second Addend = Sum – First Addend — 2 + ? = 5 — 5 - 2 = 3

 # 1

	a	b	c	d	e
Addend	73		41		9
Addend		24	40	56	
Sum	81	81		81	81

 # 2

Addend	72		40		8
Addend		25	41	57	
Sum	81	81		81	81

 # 3

Addend	71		39		7
Addend		26	42	58	
Sum	81	81		81	81

 # 4

Addend	70		38		6
Addend		27	43	59	
Sum	81	81		81	81

 # 5

Addend	69		37		5
Addend		28	44	60	
Sum	81	81		81	81

 # 6

				d	
Addend	68		36	?	4
Addend		29	45	61	
Sum	81	81		81	81

d) ? + 61 = 81

 # 7

					e
Addend	67		35		3
Addend		30	46	62	?
Sum	81	81		81	81

e) 3 + ? = 81

 # 8

Addend	66	?	34	?	2
Addend	?	31	47	63	?
Sum	81	81	?	81	81

Grade 3 Chapter 4 Lesson 5
Number 81 Complete the tables

Minuend – Subtrahend = Difference 5 – 2 = 3

Minuend = Subtrahend + Difference ? - 2 = 3 3 + 2 = 5

Subtrahend = Minuend – Difference 5 - ? = 3 5 – 3 = 2

9

	a	b	c	d	e
Minuend	81	81		81	81
Subtrahend		16	32		64
Difference	81		49	33	

10

Minuend	81	81		81	81
Subtrahend		17	33		65
Difference	80		48	32	

11

Minuend	81	81		81	81
Subtrahend		18	34		66
Difference	79		47	31	

12

Minuend	81	81		81	81
Subtrahend		19	35		67
Difference	78		46	30	

13

Minuend	81	81		81	81
Subtrahend		20	36		68
Difference	77		45	29	

14

c) ? – 37 = 44

			c		
Minuend	81	81	?	81	81
Subtrahend		21	37		69
Difference	76		44	28	

15

d) 81 – ? = 27

				d	
Minuend	81	81		81	81
Subtrahend		22	38	?	70
Difference	75		43	27	

16

Minuend	81	81	?	81	81
Subtrahend	?	23	39	?	71
Difference	74	?	42	26	?

Grade 3 Chapter 4 Lesson 6
Number 82 Complete the tables

Addend + Addend = Sum	**2 + 3 = 5**	
First Addend = Sum – Second Addend	**? + 3 = 5**	**5 – 3 = 2**
Second Addend = Sum – First Addend	**2 + ? = 5**	**5 – 2 = 3**

 # 1

	a	b	c	d	e
Addend	74		42		10
Addend		24	40	56	
Sum	82	82		82	82

 # 2

Addend	73		41		9
Addend		25	41	57	
Sum	82	82		82	82

 # 3

Addend	72		40		8
Addend		26	42	58	
Sum	82	82		82	82

 # 4

Addend	71		39		7
Addend		27	43	59	
Sum	82	82		82	82

 # 5

Addend	70		38		6
Addend		28	44	60	
Sum	82	82		82	82

 # 6

	a				
Addend	69		37		5
Addend	?	29	45	61	
Sum	82	82		82	82

a) 69 + ? = 82

 # 7

		b			
Addend	68	?	36		4
Addend		30	46	62	
Sum	82	82		82	82

b) ? + 30 = 82

 # 8

Addend	67	?	35	?	3
Addend	?	31	47	63	?
Sum	82	82	?	82	82

Grade 3 Chapter 4 Lesson 6
Number 82 Complete the tables

Minuend – Subtrahend = Difference 5 – 2 = 3
Minuend = Subtrahend + Difference ? - 2 = 3 3 + 2 = 5
Subtrahend = Minuend – Difference 5 - ? = 3 5 – 3 = 2

 # 9

	a	b	c	d	e
Minuend	82	82		82	82
Subtrahend		16	32		64
Difference	82		50	34	

 # 10

Minuend	82	82		82	82
Subtrahend		17	33		65
Difference	81		49	33	

 # 11

Minuend	82	82		82	82
Subtrahend		18	34		66
Difference	80		48	32	

 # 12

Minuend	82	82		82	82
Subtrahend		19	35		67
Difference	79		47	31	

 # 13

Minuend	82	82		82	82
Subtrahend		20	36		68
Difference	78		46	30	

 # 14 a

a) 82 - ? = 77

Minuend	82	82		82	82
Subtrahend	?	21	37		69
Difference	77		45	29	

 # 15 c

c) ? – 38 = 44

Minuend	82	82	?	82	82
Subtrahend		22	38		70
Difference	76		44	28	

 # 16

Minuend	82	82	?	82	82
Subtrahend	?	23	39	?	71
Difference	75	?	43	27	?

Grade 3 Chapter 4 Lesson 7
Number 83 Complete the tables

Addend + Addend = Sum **2 + 3 = 5**

First Addend = Sum – Second Addend **? + 3 = 5** **5 – 3 = 2**

Second Addend = Sum – First Addend **2 + ? = 5** **5 – 2 = 3**

 # 1

	a	b	c	d	e
Addend	75		43		11
Addend		24	40	56	
Sum	83	83		83	83

 # 2

Addend	74		42		10
Addend		25	41	57	
Sum	83	83		83	83

 # 3

Addend	73		41		9
Addend		26	42	58	
Sum	83	83		83	83

 # 4

Addend	72		40		8
Addend		27	43	59	
Sum	83	83		83	83

 # 5

Addend	71		39		7
Addend		28	44	60	
Sum	83	83		83	83

 # 6 d

d) ? + 61 = 83

Addend	70		38	?	6
Addend		29	45	61	
Sum	83	83		83	83

 # 7 e

e) 5 + ? = 83

Addend	69		37		5
Addend		30	46	62	?
Sum	83	83		83	83

 # 8

Addend	68	?	36	?	4
Addend	?	31	47	63	?
Sum	83	83	?	83	83

Grade 3 Chapter 4 Lesson 7
Number 83 Complete the tables

Minuend – Subtrahend = Difference 5 – 2 = 3

Minuend = Subtrahend + Difference ? - 2 = 3 3 + 2 = 5

Subtrahend = Minuend – Difference 5 - ? = 3 5 – 3 = 2

 # 9

	a	b	c	d	e
Minuend	83	83		83	83
Subtrahend		16	32		64
Difference	83		51	35	

 # 10

Minuend	83	83		83	83
Subtrahend		17	33		65
Difference	82		50	34	

 # 11

Minuend	83	83		83	83
Subtrahend		18	34		66
Difference	81		49	33	

 # 12

Minuend	83	83		83	83
Subtrahend		19	35		67
Difference	80		48	32	

 # 13

Minuend	83	83		83	83
Subtrahend		20	36		68
Difference	79		47	31	

 # 14

			c		
Minuend	83	83	?	83	83
Subtrahend		21	37		69
Difference	78		46	30	

c) ? – 37 = 46

 # 15

				d	
Minuend	83	83		83	83
Subtrahend		22	38	?	70
Difference	77		45	29	

d) 83 – ? = 29

 # 16

Minuend	83	83	?	83	83
Subtrahend	?	23	39	?	71
Difference	76	?	44	28	?

Grade 3 Chapter 4 Lesson 8
Number 84 Complete the tables

Addend + Addend = Sum **2 + 3 = 5**

First Addend = Sum – Second Addend **? + 3 = 5** **5 – 3 = 2**

Second Addend = Sum – First Addend **2 + ? = 5** **5 – 2 = 3**

 # 1

	a	b	c	d	e
Addend	76		44		12
Addend		24	40	56	
Sum	84	84		84	84

 # 2

Addend	75		43		11
Addend		25	41	57	
Sum	84	84		84	84

 # 3

Addend	74		42		10
Addend		26	42	58	
Sum	84	84		84	84

 # 4

Addend	73		41		9
Addend		27	43	59	
Sum	84	84		84	84

 # 5

Addend	72		40		8
Addend		28	44	60	
Sum	84	84		84	84

 # 6 **a** **a) 71 + ? = 84**

Addend	71		39		7
Addend	?	29	45	61	
Sum	84	84		84	84

 # 7 **b** **b) ? + 30 = 84**

Addend	70	?	38		6
Addend		30	46	62	
Sum	84	84		84	84

 # 8

Addend	69	?	37	?	5
Addend	?	31	47	63	?
Sum	84	84	?	84	84

Grade 3 Chapter 4 Lesson 8
Number 84 Complete the tables

Minuend – Subtrahend = Difference 5 – 2 = 3

Minuend = Subtrahend + Difference ? - 2 = 3 3 + 2 = 5

Subtrahend = Minuend – Difference 5 - ? = 3 5 – 3 = 2

9

	a	b	c	d	e
Minuend	84	84		84	84
Subtrahend		16	32		64
Difference	84		52	36	

10

Minuend	84	84		84	84
Subtrahend		17	33		65
Difference	83		51	35	

11

Minuend	84	84		84	84
Subtrahend		18	34		66
Difference	82		50	34	

12

Minuend	84	84		84	84
Subtrahend		19	35		67
Difference	81		49	33	

13

Minuend	84	84		84	84
Subtrahend		20	36		68
Difference	80		48	32	

14 a

a) 84 - ? = 79

Minuend	84	84		84	84
Subtrahend	?	21	37		69
Difference	79		47	31	

15 c

c) ? - 38 = 46

Minuend	84	84	?	84	84
Subtrahend		22	38		70
Difference	78		46	30	

16

Minuend	84	84	?	84	84
Subtrahend	?	23	39	?	71
Difference	77	?	45	29	?

Grade 3 Chapter 4 Lesson 9
Number 85 Complete the tables

Addend + Addend = Sum 2 + 3 = 5

First Addend = Sum – Second Addend ? + 3 = 5 5 – 3 = 2

Second Addend = Sum – First Addend 2 + ? = 5 5 – 2 = 3

 # 1

	a	b	c	d	e
Addend	77		45		13
Addend		24	40	56	
Sum	85	85		85	85

 # 2

Addend	76		44		12
Addend		25	41	57	
Sum	85	85		85	85

 # 3

Addend	75		43		11
Addend		26	42	58	
Sum	85	85		85	85

 # 4

Addend	74		42		10
Addend		27	43	59	
Sum	85	85		85	85

 # 5

Addend	73		41		9
Addend		28	44	60	
Sum	85	85		85	85

 # 6 d

d) ? + 61 = 85

Addend	72		40	?	8
Addend		29	45	61	
Sum	85	85		85	85

 # 7 e

e) 7 + ? = 85

Addend	71		39		7
Addend		30	46	62	?
Sum	85	85		85	85

 # 8

Addend	70	?	38	?	6
Addend	?	31	47	63	?
Sum	85	85	?	85	85

Grade 3 Chapter 4 Lesson 9
Number 85 Complete the tables

Minuend – Subtrahend = Difference   **5 – 2 = 3**

Minuend = Subtrahend + Difference   **? - 2 = 3**   **3 + 2 = 5**

Subtrahend = Minuend – Difference   **5 - ? = 3**   **5 – 3 = 2**

 # 9

	a	b	c	d	e
Minuend	85	85		85	85
Subtrahend		16	32		64
Difference	85		53	37	

 # 10

Minuend	85	85		85	85
Subtrahend		17	33		65
Difference	84		52	36	

 # 11

Minuend	85	85		85	85
Subtrahend		18	34		66
Difference	83		51	35	

 # 12

Minuend	85	85		85	85
Subtrahend		19	35		67
Difference	82		50	34	

 # 13

Minuend	85	85		85	85
Subtrahend		20	36		68
Difference	81		49	33	

 # 14   **c) ? – 37 = 48**

			c		
Minuend	85	85	?	85	85
Subtrahend		21	37		69
Difference	80		48	32	

 # 15   **d) 85 - ? = 31**

				d	
Minuend	85	85		85	85
Subtrahend		22	38	?	70
Difference	79		47	31	

 # 16

Minuend	85	85	?	85	85
Subtrahend	?	23	39	?	71
Difference	78	?	46	30	?

Grade 3 Chapter 4 Lesson 10
Number 86 Complete the tables

Addend + Addend = Sum **2 + 3 = 5**

First Addend = Sum – Second Addend **? + 3 = 5** **5 – 3 = 2**

Second Addend = Sum – First Addend **2 + ? = 5** **5 – 2 = 3**

 # 1

	a	b	c	d	e
Addend	78		46		14
Addend		24	40	56	
Sum	86	86		86	86

 # 2

Addend	77		45		13
Addend		25	41	57	
Sum	86	86		86	86

 # 3

Addend	76		44		12
Addend		26	42	58	
Sum	86	86		86	86

 # 4

Addend	75		43		11
Addend		27	43	59	
Sum	86	86		86	86

 # 5

Addend	74		42		10
Addend		28	44	60	
Sum	86	86		86	86

 # 6 **a** **a) 73 + ? = 86**

Addend	73		41		9
Addend	?	29	45	61	
Sum	86	86		86	86

 # 7 **b** **b) ? + 30 = 86**

Addend	72	?	40		8
Addend		30	46	62	
Sum	86	86		86	86

 # 8

Addend	71	?	39	?	7
Addend	?	31	47	63	?
Sum	86	86	?	86	86

Grade 3 Chapter 4 Lesson 10
Number 86 Complete the tables

Minuend – Subtrahend = Difference 5 – 2 = 3
Minuend = Subtrahend + Difference ? - 2 = 3 3 + 2 = 5
Subtrahend = Minuend – Difference 5 - ? = 3 5 – 3 = 2

 # 9

	a	b	c	d	e
Minuend	86	86		86	86
Subtrahend		16	32		64
Difference	86		54	38	

 # 10

Minuend	86	86		86	86
Subtrahend		17	33		65
Difference	85		53	37	

 # 11

Minuend	86	86		86	86
Subtrahend		18	34		66
Difference	84		52	36	

 # 12

Minuend	86	86		86	86
Subtrahend		19	35		67
Difference	83		51	35	

 # 13

Minuend	86	86		86	86
Subtrahend		20	36		68
Difference	82		50	34	

 # 14 a

a) 86 - ? = 81

Minuend	86	86		86	86
Subtrahend	?	21	37		69
Difference	81		49	33	

 # 15 c

c) ? - 38 = 48

Minuend	86	86	?	86	86
Subtrahend		22	38		70
Difference	80		48	32	

 # 16

Minuend	86	86	?	86	86
Subtrahend	?	23	39	?	71
Difference	79	?	47	31	?

Grade 3 Chapter 4 Lesson 11
Number 87 Complete the tables

Addend + Addend = Sum 2 + 3 = 5

First Addend = Sum – Second Addend ? + 3 = 5 5 – 3 = 2

Second Addend = Sum – First Addend 2 + ? = 5 5 – 2 = 3

 # 1

	a	b	c	d	e
Addend	79		47		15
Addend		24	40	56	
Sum	87	87		87	87

 # 2

Addend	78		46		14
Addend		25	41	57	
Sum	87	87		87	87

 # 3

Addend	77		45		13
Addend		26	42	58	
Sum	87	87		87	87

 # 4

Addend	76		44		12
Addend		27	43	59	
Sum	87	87		87	87

 # 5

Addend	75		43		11
Addend		28	44	60	
Sum	87	87		87	87

 # 6

d) ? + 61 = 87

				d	
Addend	74		42	?	10
Addend		29	45	61	
Sum	87	87		87	87

 # 7

e) 9 + ? = 87

					e
Addend	73		41		9
Addend		30	46	62	?
Sum	87	87		87	87

 # 8

Addend	72	?	40	?	8
Addend	?	31	47	63	?
Sum	87	87	?	87	87

Grade 3 Chapter 4 Lesson 11
Number 87 Complete the tables

Minuend – Subtrahend = Difference 5 – 2 = 3

Minuend = Subtrahend + Difference ? - 2 = 3 3 + 2 = 5

Subtrahend = Minuend – Difference 5 - ? = 3 5 – 3 = 2

 # 9

	a	b	c	d	e
Minuend	87	87		87	87
Subtrahend		16	32		64
Difference	87		55	39	

 # 10

Minuend	87	87		87	87
Subtrahend		17	33		65
Difference	86		54	38	

 # 11

Minuend	87	87		87	87
Subtrahend		18	34		66
Difference	85		53	37	

 # 12

Minuend	87	87		87	87
Subtrahend		19	35		67
Difference	84		52	36	

 # 13

Minuend	87	87		87	87
Subtrahend		20	36		68
Difference	83		51	35	

 # 14

			c		
Minuend	87	87	?	87	87
Subtrahend		21	37		69
Difference	82		50	34	

c) ? - 37 = 50

 # 15

				d	
Minuend	87	87		87	87
Subtrahend		22	38	?	70
Difference	81		49	33	

d) 87 – ? = 33

 # 16

Minuend	87	87	?	87	87
Subtrahend	?	23	39	?	71
Difference	80	?	48	32	?

Grade 3 Chapter 4 Lesson 12
Number 88 Complete the tables

Addend + Addend = Sum 2 + 3 = 5

First Addend = Sum – Second Addend ? + 3 = 5 5 – 3 = 2

Second Addend = Sum – First Addend 2 + ? = 5 5 – 2 = 3

 # 1

	a	b	c	d	e
Addend	80		48		16
Addend		24	40	56	
Sum	88	88		88	88

 # 2

Addend	79		47		15
Addend		25	41	57	
Sum	88	88		88	88

 # 3

Addend	78		46		14
Addend		26	42	58	
Sum	88	88		88	88

 # 4

Addend	77		45		13
Addend		27	43	59	
Sum	88	88		88	88

 # 5

Addend	76		44		12
Addend		28	44	60	
Sum	88	88		88	88

 # 6

	a				
Addend	75		43		11
Addend	?	29	45	61	
Sum	88	88		88	88

a) 75 + ? = 88

 # 7

		b			
Addend	74	?	42		10
Addend		30	46	62	
Sum	88	88		88	88

b) ? + 30 = 88

 # 8

Addend	73	?	41	?	9
Addend	?	31	47	63	?
Sum	88	88	?	88	88

Grade 3 Chapter 4 Lesson 12
Number 88 Complete the tables

Minuend – Subtrahend = Difference 5 – 2 = 3
Minuend = Subtrahend + Difference ? - 2 = 3 3 + 2 = 5
Subtrahend = Minuend – Difference 5 - ? = 3 5 – 3 = 2

 # 9

	a	b	c	d	e
Minuend	88	88		88	88
Subtrahend		16	32		64
Difference	88		56	40	

 # 10

Minuend	88	88		88	88
Subtrahend		17	33		65
Difference	87		55	39	

 # 11

Minuend	88	88		88	88
Subtrahend		18	34		66
Difference	86		54	38	

 # 12

Minuend	88	88		88	88
Subtrahend		19	35		67
Difference	85		53	37	

 # 13

Minuend	88	88		88	88
Subtrahend		20	36		68
Difference	84		52	36	

 # 14 a

a) 88 - ? = 83

Minuend	88	88		88	88
Subtrahend	?	21	37		69
Difference	83		51	35	

 # 15 c

c) ? – 38 = 50

Minuend	88	88	?	88	88
Subtrahend		22	38		70
Difference	82		50	34	

 # 16

Minuend	88	88	?	88	88
Subtrahend	?	23	39	?	71
Difference	81	?	49	33	?

Grade 3 Chapter 4 Lesson 13
Number 89 Complete the tables

Addend + Addend = Sum — **2 + 3 = 5**

First Addend = Sum – Second Addend — **? + 3 = 5** **5 – 3 = 2**

Second Addend = Sum – First Addend — **2 + ? = 5** **5 – 2 = 3**

 # 1

	a	b	c	d	e
Addend	81		49		17
Addend		24	40	56	
Sum	89	89		89	89

 # 2

Addend	80		48		16
Addend		25	41	57	
Sum	89	89		89	89

 # 3

Addend	79		47		15
Addend		26	42	58	
Sum	89	89		89	89

 # 4

Addend	78		46		14
Addend		27	43	59	
Sum	89	89		89	89

 # 5

Addend	77		45		13
Addend		28	44	60	
Sum	89	89		89	89

 # 6 — d — **d) ? + 61 = 89**

Addend	76		44	?	12
Addend		29	45	61	
Sum	89	89		89	89

 # 7 — e — **e) 11 + ? = 89**

Addend	75		43		11
Addend		30	46	62	?
Sum	89	89		89	89

 # 8

Addend	74	?	42	?	10
Addend	?	31	47	63	?
Sum	89	89	?	89	89

Grade 3 Chapter 4 Lesson 13
Number 89 Complete the tables

Minuend – Subtrahend = Difference 5 – 2 = 3
Minuend = Subtrahend + Difference ? - 2 = 3 3 + 2 = 5
Subtrahend = Minuend – Difference 5 - ? = 3 5 – 3 = 2

 # 9

	a	b	c	d	e
Minuend	89	89		89	89
Subtrahend		16	32		64
Difference	89		57	41	

 # 10

Minuend	89	89		89	89
Subtrahend		17	33		65
Difference	88		56	40	

 # 11

Minuend	89	89		89	89
Subtrahend		18	34		66
Difference	87		55	39	

 # 12

Minuend	89	89		89	89
Subtrahend		19	35		67
Difference	86		54	38	

 # 13

Minuend	89	89		89	89
Subtrahend		20	36		68
Difference	85		53	37	

 # 14 c) ? – 37 = 52

			c		
Minuend	89	89	?	89	89
Subtrahend		21	37		69
Difference	84		52	36	

 # 15 d) 89 - ? = 35

				d	
Minuend	89	89		89	89
Subtrahend		22	38	?	70
Difference	83		51	35	

 # 16

Minuend	89	89	?	89	89
Subtrahend	?	23	39	?	71
Difference	82	?	50	34	?

Grade 3 Chapter 4 Lesson 14
Review 4 A Numbers 80 - 84
Complete the tables

# 1	a	b	c	d	e
Addend	65		33		1
Addend		31	47	63	
Sum	80	80		80	80

2

Minuend	80	80		80	80
Subtrahend		31	47		79
Difference	65		33	17	

3

Addend	66		34		2
Addend		31	47	63	
Sum	81	81		81	81

4

Minuend	81	81	?	81	81
Subtrahend	?	31	47	?	79
Difference	66	?	34	18	?

5

Addend	67	?	35	?	3
Addend	?	31	47	63	?
Sum	82	82	?	82	82

Grade 3 Chapter 4 Lesson 14

Review 4 B Numbers 80 - 84

Complete the tables

# 6	a	b	c	d	e
Minuend	82	82		82	82
Subtrahend		31	47		79
Difference	67		35	19	

7

Addend	68		36		4
Addend		31	47	63	
Sum	83	83		83	83

8

Minuend	83	83		83	83
Subtrahend		31	47		79
Difference	68		36	20	

9

Addend	69	?	37	?	5
Addend	?	31	47	63	?
Sum	84	84	?	84	84

10

Minuend	84	84	?	84	84
Subtrahend	?	31	47	?	79
Difference	69	?	37	21	?

Grade 3 Chapter 4 Lesson 15

Test 4 A Numbers 85 - 89

Complete the tables

1

	a	b	c	d	e
Addend	70		38		6
Addend		31	47	63	
Sum	85	85		85	85

2

Minuend	85	85		85	85
Subtrahend		31	47		79
Difference	70		38	22	

3

Addend	71		39		7
Addend		31	47	63	
Sum	86	86		86	86

4

Minuend	86	86	?	86	86
Subtrahend	?	31	47	?	79
Difference	71	?	39	23	?

5

Addend	72	?	40	?	8
Addend	?	31	47	63	?
Sum	87	87	?	87	87

Grade 3 Chapter 4 Lesson 15

Test 4 B Numbers 85 - 89

Complete the tables

6

	a	b	c	d	e
Minuend	87	87		87	87
Subtrahend		31	47		79
Difference	72		40	24	

7

Addend	73		41		9
Addend		31	47	63	
Sum	88	88		88	88

8

Minuend	88	88		88	88
Subtrahend		31	47		79
Difference	73		41	25	

9

Addend	74	?	42	?	10
Addend	?	31	47	63	?
Sum	89	89	?	89	89

10

Minuend	89	89	?	89	89
Subtrahend	?	31	47	?	79
Difference	74	?	42	26	?

Grade 3 Chapter 4 Lesson 16

Review 1 A

Put correct number instead questions mark

 # 1 ? + 19 = 50
 # 2 22 + ? = 51

 # 3 ? + 29 = 52
 # 4 24 + ? = 53

 # 5 ? + 39 = 54
 # 6 ? – 19 = 31

 # 7 51 - ? = 29
 # 8 ? – 39 = 13

 # 9 53 - ? = 29
 # 10 ? – 29 = 25

Complete the tables

	#11	#12	#13	#14	# 15
Addend		29	29		19
Addend	29		28	19	?
Sum	55	56		58	59

15) 19 + ? = 59

	#16	#17	#18	#19	#20
Minuend		56	57		59
Subtrahend	26		29	19	?
Difference	29	27		39	19

20) 59 - ? = 19

Grade 3 Chapter 4 Lesson 16

Review 1 B

Put correct number instead question mark

1 ? + 31 = 50
2 23 + ? = 51

3 ? + 13 = 52
4 25 + ? = 53

5 ? + 29 = 54
6 ? – 29 = 21

7 51 - ? = 39
8 ? – 29 = 23

9 53 - ? = 27
10 ? – 28 = 26

Complete the tables

	#11	#12	#13	#14	# 15
Addend		17	28		40
Addend	26		29	29	?
Sum	55	56		58	59

15) 40 + ? = 59

	#16	#17	#18	#19	#20
Minuend		56	57		59
Subtrahend	27		29	29	?
Difference	28	29		29	29

20) 59 - ? = 29

Grade 3 Chapter 4 Lesson 17

Review 1 A

Find correct number instead question mark

1 ? + 19 = 80
2 42 + ? = 81
3 ? + 49 = 82
4 24 + ? = 83
5 ? + 69 = 84
6 ? – 19 = 61
7 81 - ? = 39
8 ? – 49 = 33
9 83 - ? = 59
10 ? – 69 = 15

Complete the tables

	#11	#12	#13	#14	# 15
Addend	?	37	48		60
Addend	59		39	29	?
Sum	85	86	?	88	89

	#16	#17	#18	#19	#20
Minuend		86	87	?	89
Subtrahend	26	?	48	29	
Difference	59	37		59	29

Review 1 B

Find correct number instead question mark

1 ? + 19 = 90
2 52 + ? = 91
3 ? + 59 = 92
4 34 + ? = 93
5 ? + 69 = 94
6 ? – 39 = 51
7 91 - ? = 49
8 ? – 49 = 43
9 93 - ? = 59
10 ? – 59 = 35

Complete the tables

15) 70 + ? = 99

	#11	#12	#13	#14	# 15
Addend		47	58		70
Addend	59		39	39	?
Sum	95	96		98	99

20) 99 - ? = 39

	#16	#17	#18	#19	#20
Minuend		96	97		99
Subtrahend	36		59	39	?
Difference	59	57		59	39

Grade 3 Chapter 4 Lesson 18
Review 3 A (Basic Problems)
Add or Subtract numbers in column

# 1	53 + 37	# 6	93 -58
# 2	69 +25	# 7	97 -39
# 3	69 +29	# 8	91 -67
# 4	56 +36	# 9	95 - 47
# 5	57 +39	# 10	99 - 59

Add or Subtract numbers in row

11) 73 + 18
12) 74 + 19

13) 66 + 29
14) 79 + 18

15) 69 + 21
16) 90 – 71

17) 92 – 64
18) 94 – 49

19) 96 – 57
20) 93 – 37

Grade 3 Chapter 4 Lesson 18
Review 3 B (Basic Problems)
Add or Subtract numbers in column

# 1	67 + 24	# 6	92 - 58
# 2	67 +28	# 7	96 - 39
# 3	68 +31	# 8	90 -65
# 4	59 +34	# 9	94 - 57
# 5	58 +39	# 10	98 - 29

Add or Subtract numbers in row

11) 29 + 61
12) 59 + 33

13) 49 + 45
14) 67 + 29

15) 66 + 29
16) 91 – 29

17) 93 – 34
18) 95 – 29

19) 97 – 48
20) 97 – 41

Grade 3 Chapter 4 Lesson 19

Final Exam 3 A (Basic Problems)

Add or Subtract Numbers in Row

# 1	31 + 59
# 2	69 + 22
# 3	13 + 79
# 4	69 + 24
# 5	35 + 59
# 6	90 – 31
# 7	91 – 69
# 8	92 – 13
# 9	93 – 69
# 10	94 – 35

Add or Subtract Numbers in Column

# 11 46 +49	# 12 39 +57	# 13 68 + 29	# 14 19 +79	# 15 80 + 19
# 16 95 -49	# 17 96 -57	# 18 97 -29	# 19 98 -79	# 20 99 - 19

Grade 3 Chapter 4 Lesson 19
Final Exam 3 B (Basic Problems)

Add or Subtract Numbers in Row

# 1	41 + 49
# 2	59 + 32
# 3	23 + 69
# 4	59 + 34
# 5	25 + 69
# 6	90 – 41
# 7	91 – 59
# 8	92 – 23
# 9	93 – 59
# 10	94 – 45

Add or Subtract Numbers in Column

# 11 36 +59	# 12 49 +47	# 13 58 + 39	# 14 29 +69	# 15 70 + 29
# 16 95 -59	# 17 96 -67	# 18 97 -39	# 19 98 -69	# 20 99 - 29

 # 1 – 20 odd or # 1, 3, 5, 7, 9, 11, 13, 15, 17, 19
Grading Scale: A =10 points, B=9points, C=8 points, D=7 points, F=0-6 points

Final Exam 3 B

# 1	90	# 11	95	# 6	49	# 16	36
# 2	91	# 12	96	# 7	32	# 17	29
# 3	92	# 13	97	# 8	69	# 18	58
# 4	93	# 14	98	# 9	34	# 19	29
# 5	94	# 15	99	# 10	49	# 20	70

Grade 3 Answers Chapter 4

Lesson 1 Summary 1

# 1	30	# 11	8	# 21	70	# 31	48
# 2	31	# 12	19	# 22	72	# 32	46
# 3	34	# 13	7	# 23	74	# 33	39
# 4	35	# 14	16	# 24	76	# 34	29
# 5	36	# 15	27	# 25	78	# 35	30
# 6	33	# 16	19	# 26	71	# 36	49
# 7	32	# 17	8	# 27	73	# 37	47
# 8	39	# 18	11	# 28	75	# 38	45
# 9	38	# 19	19	# 29	77	# 39	38
# 10	37	# 20	19	# 30	79	# 40	19

Lesson 1 Summary 2

# 1	40	# 11	13	# 21	80	# 31	14
# 2	42	# 12	14	# 22	82	# 32	25
# 3	44	# 13	29	# 23	84	# 33	38
# 4	46	# 14	28	# 24	86	# 34	48
# 5	48	# 15	34	# 25	88	# 35	30
# 6	31	# 16	25	# 26	81	# 36	15
# 7	43	# 17	16	# 27	83	# 37	24
# 8	45	# 18	27	# 28	85	# 38	37
# 9	47	# 19	28	# 29	87	# 39	47
# 10	49	# 20	19	# 30	89	# 40	59

Lesson 2 Summary 3

# 1	6	# 11	14	# 21	21	# 31	33
# 2	17	# 12	21	# 22	25	# 32	37
# 3	5	# 13	8	# 23	29	# 33	26
# 4	8	# 14	25	# 24	37	# 34	28
# 5	18	# 15	26	# 25	39	# 35	20
# 6	4	# 16	8	# 26	38	# 36	60
# 7	18	# 17	22	# 27	36	# 37	72
# 8	5	# 18	16	# 28	39	# 38	64
# 9	9	# 19	38	# 29	18	# 39	66
# 10	9	# 20	18	# 30	40	# 40	68

Grade 3 Answers Chapter 4

Lesson 2 Summary 4

# 1	35	# 11	58	# 21	18	# 31	90
# 2	36	# 12	18	# 22	74	# 32	28
# 3	54	# 13	37	# 23	95	# 33	62
# 4	38	# 14	52	# 24	22	# 34	93
# 5	39	# 15	17	# 25	61	# 35	45
# 6	25	# 16	59	# 26	92	# 36	39
# 7	37	# 17	29	# 27	45	# 37	95
# 8	55	# 18	27	# 28	29	# 38	48
# 9	28	# 19	53	# 29	99	# 39	58
# 10	40	# 20	22	# 30	41	# 40	99

Lesson 3 Review 1 A

	a	b	c	d		a	b	c	d
# 1	28	1	30	13	# 2	2	29	30	16
# 3	28	2	30	0	# 4	28	2	30	0
# 5	4	29	2	31	# 6	28	3	30	1
# 7	5	29	3	31	# 8	28	4	30	2
# 9	6	29	4	31	# 10	28	5	30	3

Lesson 3 Review 1 B

	a	b	c	d		a	b	c	d
# 11	35	29	5	31	# 12	35	6	30	4
# 13	36	29	6	31	# 14	36	29	6	31
# 15	37	29	7	31	# 16	37	29	7	31
# 17	38	29	8	31	# 18	38	29	8	31
# 19	39	29	9	31	# 20	39	29	9	31

Lesson 4 Number 80

	a b c d e		a b c d e
# 1	8, 56, 80, 24, 72	# 9	0, 64, 80, 48, 16
# 2	9, 55, 80, 23, 73	# 10	1, 63, 80, 49, 15
# 3	10, 54, 80, 22, 74	# 11	2, 62, 80, 50, 14
# 4	11, 53, 80, 21, 75	# 12	3, 61, 80, 51, 13
# 5	12, 52, 80, 20, 76	# 13	4, 60, 80, 52, 12
# 6	13, 51, 80, 19, 77	# 14	5, 59, 80, 53, 11
# 7	14, 50, 80, 18, 78	# 15	6, 58, 80, 54, 10
# 8	15, 49, 80, 17, 79	# 16	7, 57, 80, 55, 9

Grade 3 Answers Chapter 4

Lesson 5 Number 81

	a b c d e		a b c d e
# 1	8, 57, 81, 25, 72	# 9	0, 65, 81, 48, 17
# 2	9, 56, 81, 24, 73	# 10	1, 64, 81, 49, 16
# 3	10, 55, 81, 23, 74	# 11	2, 63, 81, 50, 15
# 4	11, 54, 81, 22, 75	# 12	3, 62, 81, 51, 14
# 5	12, 53, 81, 21, 76	# 13	4, 61, 81, 52, 13
# 6	13, 52, 81, 20, 77	# 14	5, 60, 81, 53, 12
# 7	14, 51, 81, 19, 78	# 15	6, 59, 81, 54, 11
# 8	15, 50, 81, 18, 79	# 16	7, 58, 81, 55, 10

Lesson 6 Number 82

	a b c d e		a b c d e
# 1	8, 58, 82, 26, 72	# 9	0, 66, 82, 48, 18
# 2	9, 57, 82, 25, 73	# 10	1, 65, 82, 49, 17
# 3	10, 56, 82, 24, 74	# 11	2, 64, 82, 50, 16
# 4	11, 55, 82, 23, 75	# 12	3, 63, 82, 51, 15
# 5	12, 54, 82, 22, 76	# 13	4, 62, 82, 52, 14
# 6	13, 53, 82, 21, 77	# 14	5, 61, 82, 53, 13
# 7	14, 52, 82, 20, 78	# 15	6, 60, 82, 54, 12
# 8	15, 51, 82, 19, 79	# 16	7, 59, 82, 55, 11

Lesson 7 Number 83

# 1	8, 59, 83, 27, 72	# 9	0, 67, 83, 48, 19
# 2	9, 58, 83, 26, 73	# 10	1, 66, 83, 49, 18
# 3	10, 57, 83, 25, 74	# 11	2, 65, 83, 50, 17
# 4	11, 56, 83, 24, 75	# 12	3, 64, 83, 51, 16
# 5	12, 55, 83, 23, 76	# 13	4, 63, 83, 52, 15
# 6	13, 54, 83, 22, 77	# 14	5, 62, 83, 53, 14
# 7	14, 53, 83, 21, 78	# 15	6, 61, 83, 54, 13
# 8	15, 52, 83, 20, 79	# 16	7, 60, 83, 55, 12

Lesson 8 Number 84

	a b c d e		a b c d e
# 1	8, 60, 84, 28, 72	# 9	0, 68, 84, 48, 20
# 2	9, 59, 84, 27, 73	# 10	1, 67, 84, 49, 19
# 3	10, 58, 84, 26, 74	# 11	2, 66, 84, 50, 18
# 4	11, 57, 84, 25, 75	# 12	3, 65, 84, 51, 17
# 5	12, 56, 84, 24, 76	# 13	4, 64, 84, 52, 16
# 6	13, 55, 84, 23, 77	# 14	5, 63, 84, 53, 15
# 7	14, 54, 84, 22, 78	# 15	6, 62, 84, 54, 14
# 8	15, 53, 84, 21, 79	# 16	7, 61, 84, 55, 13

Grade 3 Answers Chapter 4

Lesson 9 Number 85

	a b c d e		a b c d e
# 1	8, 61, 85, 29, 72	# 9	0, 69, 85, 48, 21
# 2	9, 60, 85, 28, 73	# 10	1, 68, 85, 49, 20
# 3	10, 59, 85, 27, 74	# 11	2, 67, 85, 50, 19
# 4	11, 58, 85, 26, 75	# 12	3, 66, 85, 51, 18
# 5	12, 57, 85, 25, 76	# 13	4, 65, 85, 52, 17
# 6	13, 56, 85, 24, 77	# 14	5, 64, 85, 53, 16
# 7	14, 55, 85, 23, 78	# 15	6, 63, 85, 54, 15
# 8	15, 54, 85, 22, 79	# 16	7, 62, 85, 55, 14

Lesson 10 Number 86

	a b c d e		a b c d e
# 1	8, 62, 86, 30, 72	# 9	0, 70, 86, 48, 22
# 2	9, 61, 86, 29, 73	# 10	1, 69, 86, 49, 21
# 3	10, 60, 86, 28, 74	# 11	2, 68, 86, 50, 20
# 4	11, 59, 86, 27, 75	# 12	3, 67, 86, 51, 19
# 5	12, 58, 86, 26, 76	# 13	4, 66, 86, 52, 18
# 6	13, 57, 86, 25, 77	# 14	5, 65, 86, 53, 17
# 7	14, 56, 86, 24, 78	# 15	6, 64, 86, 54, 16
# 8	15, 55, 86, 23, 79	# 16	7, 63, 86, 55, 15

Lesson 11 Number 87

# 1	8, 63, 87, 31, 72	# 9	0, 71, 87, 48, 23
# 2	9, 62, 87, 30, 73	# 10	1, 70, 87, 49, 22
# 3	10, 61, 87, 29, 74	# 11	2, 69, 87, 50, 21
# 4	11, 60, 87, 28, 75	# 12	3, 68, 87, 51, 20
# 5	12, 59, 87, 27, 76	# 13	4, 67, 87, 52, 19
# 6	13, 58, 87, 26, 77	# 14	5, 66, 87, 53, 18
# 7	14, 57, 87, 25, 78	# 15	6, 65, 87, 54, 17
# 8	15, 56, 87, 24, 79	# 16	7, 64, 87, 55, 16

Lesson 12 Number 88

	a b c d e		a b c d e
# 1	8, 64, 88, 32, 72	# 9	0, 72, 88, 48, 24
# 2	9, 63, 88, 31, 73	# 10	1, 71, 88, 49, 23
# 3	10, 62, 88, 30, 74	# 11	2, 70, 88, 50, 22
# 4	11, 61, 88, 29, 75	# 12	3, 69, 88, 51, 21
# 5	12, 60, 88, 28, 76	# 13	4, 68, 88, 52, 20
# 6	13, 59, 88, 27, 77	# 14	5, 67, 88, 53, 19
# 7	14, 58, 88, 26, 78	# 15	6, 66, 88, 54, 18
# 8	15, 57, 88, 25, 79	# 16	7, 65, 88, 55, 17

Grade 3 Answers Chapter 4

Lesson 13 Number 89

	a, b, c, d, e		a, b, c, d, e
# 1	8, 65, 89, 33, 72	# 9	0, 73, 89, 48, 25
# 2	9, 64, 89, 32, 73	# 10	1, 72, 89, 49, 24
# 3	10, 63, 89, 31, 74	# 11	2, 71, 89, 50, 23
# 4	11, 62, 89, 30, 75	# 12	3, 70, 89, 51, 22
# 5	12, 61, 89, 29, 76	# 13	4, 69, 89, 52, 21
# 6	13, 60, 89, 28, 77	# 14	5, 68, 89, 53, 20
# 7	14, 59, 89, 27, 78	# 15	6, 67, 89, 54, 19
# 8	15, 58, 89, 26, 79	# 16	7, 66, 89, 55, 18

Lesson 14 Review 4 A

	a	b	c	d	e
# 1	15	49	80	17	79
# 2	15	49	80	63	1
# 3	15	50	81	18	79
# 4	15	50	81	63	2
# 5	15	51	82	19	79

Review 4 B

# 6	15	51	82	63	3
# 7	15	52	83	20	79
# 8	15	52	83	63	4
# 9	15	53	84	21	79
# 10	15	53	84	63	5

Lesson 15 Test 4 A

	a	b	c	d	e
# 1	15	54	85	22	79
# 2	15	54	85	63	6
# 3	15	55	86	23	79
# 4	15	55	86	63	7
# 5	15	56	87	24	79

Test 4 B

# 6	15	56	87	63	8
# 7	15	57	88	25	79
# 8	15	57	88	63	9
# 9	15	58	89	26	79
# 10	15	58	89	63	10

Grade 3 Answers Chapter 4

Lesson 16 Review 1 A

# 1	31	# 11	26
# 2	29	# 12	27
# 3	23	# 13	57
# 4	29	# 14	39
# 5	15	# 15	40
# 6	50	# 16	55
# 7	22	# 17	29
# 8	52	# 18	28
# 9	24	# 19	58
# 10	54	# 20	40

Lesson 16 Review 1 B

# 1	19	# 11	29
# 2	29	# 12	39
# 3	39	# 13	57
# 4	29	# 14	29
# 5	25	# 15	19
# 6	50	# 16	55
# 7	12	# 17	27
# 8	52	# 18	28
# 9	26	# 19	58
# 10	54	# 20	30

Lesson 17

Review 1 A

# 1	61	# 11	26
# 2	39	# 12	49
# 3	33	# 13	87
# 4	59	# 14	59
# 5	15	# 15	29
# 6	80	# 16	85
# 7	42	# 17	49
# 8	82	# 18	39
# 9	24	# 19	88
# 10	84	# 20	60

Review 1 B

# 1	71	# 11	36
# 2	39	# 12	49
# 3	33	# 13	97
# 4	59	# 14	59
# 5	25	# 15	29
# 6	90	# 16	95
# 7	42	# 17	39
# 8	92	# 18	38
# 9	34	# 19	98
# 10	94	# 20	60

Grade 3 Answers Chapter 4

Lesson 18 (Basic Problems)

Review 3 A

# 1	90	# 6	35	# 11	91	# 16	19
# 2	94	# 7	58	# 12	93	# 17	28
# 3	98	# 8	24	# 13	95	# 18	45
# 4	92	# 9	48	# 14	97	# 19	39
# 5	96	#10	40	# 15	90	# 20	56

Review 3 B

# 1	91	# 6	34	# 11	90	# 16	62
# 2	95	# 7	57	# 12	92	# 17	59
# 3	99	# 8	25	# 13	94	# 18	66
# 4	93	# 9	37	# 14	96	# 19	49
# 5	97	# 10	69	# 15	95	# 20	56

Lesson 19 (Basic Problems)

Final Exam 3 A

# 1	90	# 11	95
# 2	91	# 12	96
# 3	92	# 13	97
# 4	93	# 14	98
# 5	94	# 15	99
# 6	59	# 16	46
# 7	22	# 17	39
# 8	79	# 18	68
# 9	24	# 19	19
# 10	59	# 20	80

Final Exam 3 B

# 1	90	# 11	95
# 2	91	# 12	96
# 3	92	# 13	97
# 4	93	# 14	98
# 5	94	# 15	99
# 6	49	# 16	36
# 7	32	# 17	29
# 8	69	# 18	58
# 9	34	# 19	29
# 10	49	# 20	70

Students!! Attention!!

Adults, help please students 1 – 4 grades to read and understand advice.

When I was in 5th grade I did not like math. I am not always understood math. I was 11 years girl. Next 6th grade I got a very good teacher. She explains very clearly and I understood her very well. I was worked very hard and I was not afraid to ask questions. Final Grade in Algebra in 6th grade I got an A. I said in class that I want to be a teacher of math. But some students said that I am not enough smart to be a math teacher. I said I will be work hard with math. I did it, I got a Master's degree in math and education, I love my students, I was teaching many years, in High School, part-time college math instructor, full-time university math lecturer.

Some people do a simple job, eat, sleep, go to work. People with a bad brain have not so a good job, not so good salary. If you look at TV how someone's dance, you will be not changing your brain. But if you take a dancing class this helps your body and brain ability. The same if you want to use the calculator, but the calculator not always with you. Much better if you learn how you need to do any operations with numbers without a calculator. If you read this math book you will know what you need to do. This teaches you how to improve your memory, attention, logic thinking, how to read your math books every day, your brain works better. You can work 50 minutes with math, 10 minutes break maybe dance. Do not afraid to do mistakes, fixed and move to the next step. Don't listen to anyone, choose what do you like more.

Students need to know if you work with the brain every day, your brain works better. You need a good brain in any job, in any class. When you finish High School, you need to get any profession, you need to go to college and get a good job and a better salary. You need to learn any subject, its work for the brain. Read the book, do the exercises yourself, try your best, and then check your answer at the end of the chapter. Be honest with yourself. You need to believe in yourself. The brain is like muscles. People with good brain lives longer. The professors at my university were over 80 years old, wise, and optimistic. It was a great pleasure to talk to them.

Made in the USA
Monee, IL
05 November 2026

67967403R10072